Cultural Studies of Science Education

Volume 20

Series Editors
Catherine Milne, Steinhardt School of Culture, Education, and Human
Development, New York University, New York, USA

Christina Siry, Institute of Applied Educational Sciences,
The University of Luxembourg, Esch-sur-Alzette, Luxembourg

The series is unique in focusing on the publication of scholarly works that employ social and cultural perspectives as foundations for research and other scholarly activities in the three fields implied in its title: science education, education, and social studies of science.

The aim of the series is to promote transdisciplinary approaches to scholarship in science education that address important topics in the science education including the teaching and learning of science, social studies of science, public understanding of science, science/technology and human values, science and literacy, ecojustice and science, indigenous studies and science and the role of materiality in science and science education. Cultural Studies of Science Education, the book series explicitly aims at establishing such bridges and at building new communities at the interface of currently distinct discourses. In this way, the current almost exclusive focus on science education on school learning would be expanded becoming instead a focus on science education as a cultural, cross-age, cross-class, and cross-disciplinary phenomenon.

The book series is conceived as a parallel to the journal Cultural Studies of Science Education, opening up avenues for publishing works that do not fit into the limited amount of space and topics that can be covered within the same text. Book proposals for this series may be submitted to the Publishing Editor: Claudia Acuna E-mail: Claudia.Acuna@springer.com.

More information about this series at http://www.springer.com/series/8286

Kathrin Otrel-Cass • Maria Andrée
Minjung Ryu
Editors

Examining Ethics in Contemporary Science Education Research

Being Responsive and Responsible

Springer

Editors
Kathrin Otrel-Cass
Institute of Education Research and Teacher
Education
University of Graz
Graz, Austria

Maria Andrée
Department of Mathematics and Science
Education
Stockholm University
Stockholm, Sweden

Minjung Ryu
Department of Chemistry and Learning
Sciences Research Institute
University of Illinois at Chicago
Chicago, Illinois, USA

ISSN 1879-7229   ISSN 1879-7237   (electronic)
Cultural Studies of Science Education
ISBN 978-3-030-50923-1   ISBN 978-3-030-50921-7   (eBook)
https://doi.org/10.1007/978-3-030-50921-7

This Springer imprint is published by the registered company Springer Nature Switzerland AG
The registered company address is: Gewerbestrasse 11, 6330 Cham, Switzerland

Contents

About the Editors

Kathrin Otrel-Cass is a Professor of education and digital transformations at the University of Graz, Austria. Her research interests are often of interdisciplinary in nature with focus on digital visual anthropology and variety of qualitative, ethnographic methodologies. She works with various practitioners and experts in environments where people are working with science/technology/engineering practices or their knowledge products. Her research is often set in schools but is not exclusive to those environments. Her research interest in visual ethnography has led to the establishment of a video research laboratories at Aalborg University and the University of Graz with a focus on the organized analysis of video recorded data and ethical research practices involving visual data. Kathrin is also a member of the Human Factor in Digital Transformation research network at the University of Graz.

Maria Andrée is an Associate Professor of science education at the Department of Mathematics and Science Education at Stockholm University, Sweden. Drawing primarily on socio-cultural theory, her research focuses on science education practices and the conditions for students' participation and learning, particularly in relation to questions of science curriculum, scientific literacy, and citizenship. She primarily works with ethnographic and design-based research studies in science education. She has also pursued a line of research concerning the involvement of external actors in science, technology, and mathematics education. She is currently one of the scientific leaders of *Stockholm Teaching & Learning Studies* – a platform for research in collaboration with teachers – designed to initiate, support, and conduct small-scale classroom-based didactic research on teaching and learning. She has published in peer-reviewed journals including *International Journal of Science Education*, *Research in Science Education*, *Cultural Studies of Science Education*, and *Journal of Curriculum Studies*, among others.

Minjung Ryu is an Assistant Professor of chemistry and learning sciences at the University of Illinois at Chicago, USA. Her research focuses on STEM learning and participation of cultural and linguistic minority students. Employing ethnography and discourse analysis, she examines how racial, ethnic, and linguistic minority students engage in STEM discourses using multilingual and multimodal means and what are ways to design learning environments to improve these students' learning experiences. Within this research interest, she has worked with resettled Burmese refugee teens in a community-based afterschool program in the USA where the teens learn STEM knowledge to transform their communities and global societies. She also has collaborated with high school science teachers to develop instructional materials and practices to support English learners in linguistically superdiverse classrooms. Minjung has published in *Journal of Research in Science Teaching*, *Science Education*, and *International Journal of Science Education*.

Chapter 1
Ethics in Contemporary Science Education Research

Kathrin Otrel-Cass, Maria Andrée, and Minjung Ryu

1.1 Introduction

Is there really a need for another book that discusses research ethics? Is there in fact a need to write about research ethics specific to science education researchers? Are there ethics considerations that go beyond those of educational researcher in general?

We think there is. The science education research community has greatly contributed to the growing understanding on how subject specific learning and teaching can be improved or on what is taking place already in order to highlight and unpack good practices. We have gained significant and detailed insights into what makes science difficult to learn and why we should consider that science practices represent very specific cultural practices that are not necessarily open to all. What makes science education research also unique is that the subject itself is of political interest. Together with mathematics and engineering education, science is often described as one of the subjects that can ensure a nation's economic well-being and international competitiveness in the future. Beyond 2000 (Millar and Osborne 1998), or the Relevance of Science Education study (Schreiner and Sjøberg 2004), AAAS's (American Association for the Advancement of Science) Project 2061 (AAAS 1993), as well as Osborne and Dillon's critical reflections on science education in Europe (Osborne and Dillon 2008) are just a few examples that emphasize the importance of science education for the nation's prosperity and security.

K. Otrel-Cass (✉)
University of Graz, Graz, Austria
e-mail: kathrin.otrel-cass@uni-graz.at

M. Andrée
Stockholm University, Stockholm, Sweden

M. Ryu
University of Illinois at Chicago, Chicago, IL, USA

© Springer Nature Switzerland AG 2020
K. Otrel-Cass et al. (eds.), *Examining Ethics in Contemporary Science Education Research*, Cultural Studies of Science Education 20,
https://doi.org/10.1007/978-3-030-50921-7_1

Not surprisingly, therefore, that research funding from a variety of funding bodies is available for science education research. While funding drives the proliferation of knowledge production through research to some degree, neoliberal realities that many universities face these days mean that securing funding becomes a necessity to pursue for most researchers (Leathwood and Read 2013). The neoliberal realities also mean that researchers are under pressure to produce knowledge at a fast pace, which may reduce the availability of time to reflect on the various nuances in their research practices. This research context inevitably has an impact on research ethics and requires careful ethical reflection and deliberation both at the individual level and at the community level.

What we hope to achieve with this book is to remind our fellow science education researchers of the ethical responsibilities to take care of the communities that we study and unmask traditional arguments and approaches. What warrants this conversation about research ethics, despite the plethora of existing resources for maintaining research ethics, is the changing condition of science education research that affects research practices in unique ways. Those changes may be of a technological nature, for instance through the possibilities to digitally capture data. New kinds of ethical questions that could arise here have to do with how we deal with and address the 'datafication' of our participants' lives. Changing conditions of science education also include new insights gained from different, but related, fields of study, for example neuroscience research. On one hand, the insights we have gained through years of research in these fields by themselves are changing conditions because such new insights require us to revisit our assumptions and approaches to teaching and learning science. On the other hand, we need to consider how those fields differ from science education research in the way in which researchers make sense of the information i.e. data and the potential benefits and risks such information presents to the knowledge production processes and arguments used in science education. We may also need to reconsider guidelines on ethical practices, when mobile technology that easily captures and distributes written and visual accounts of our research can also easily distribute participants' information without our knowledge (for example when our audiences take photos away from the presentations we give. A question we need to ask is how realistic it is when researchers claim (and most likely try) to ensure participants' anonymity and confidentiality of data or whether we fall foul of looking away. Even when we are withholding names, currently available technologies that allow for face or voice recognition are becoming smarter and are or will be equipped with cognitive powers that can self-operate without being prompted by human actors (Hayles 2017). In this technological context, the protection of anonymity and confidentiality faces new challenges. Pereira et al. (2014) pick up on what it means to think about 'the right to be forgotten' in the digital age (brought to the fore through a law introduced first by the European Commission in 2012). The authors emphasise that "the fragmentation of personal information dispersed across different web platforms creates vulnerabilities for our identity and other aspects of what constitutes our personality" (p.3). Data that is collected and utilised for science education research may have been digitally harvested, refabricated and reorganized, to be presented in online publications where others may capture and take away digital snapshots of people's presented identities.

The intention of this book is to reflect on contemporary challenges in science education under these changing conditions to initiate a renewed conversation in what ways we should and can adjust and refine our research practices in order to ethically move science education research forward. In the following sections, we turn to some key issues that we believe need our attention and that have been picked up in individual chapters of this book in a variety of ways: the nature of regulatory frameworks that shape our research practices; the need to develop a community responsibility in order to advance our ethical practices further; new methodological frameworks that influence our research ethics, with a special focus on visual methodologies; and particular ethical challenges relevant to science education. Finally, we conclude with an overview of the contributions to this book.

1.2 Ethical Regulations as a Minimum

Science education research involves by and large the study of people (often young and vulnerable people) and their practices in one way or another. Researchers working within this kind of humanities and social science research follow guidelines and frameworks that are often set by country specific ethics committees and shaped by research codes of conducts. Examples of regulatory frameworks across the world include the Canadian Tri-Council (https://research.ucalgary.ca/conduct-research/funding/apply-grants/external-grants/tri-council), the Australian Research Council (https://www.arc.gov.au/policies-strategies/policy/codes-and-guidelines), the European Commission (https://ec.europa.eu/research/swafs/index.cfm?pg=policy&lib=ethics), the Research Council of Norway and its National Committee for Research Ethics in the Social Sciences and Humanities (NESH 2016), the UK's NHS National Research Ethics Service (https://www.hra.nhs.uk/about-us/committees-and-services/res-and-recs/) and the Research Ethics Framework (2015) of the ESRC (Economic and Social Research Council) General Guidelines (https://esrc.ukri.org/files/funding/guidance-for-applicants/esrc-framework-for-research-ethics-2015/), the Forum for Ethical Review Committees in Asia and the Western Pacific (FERCAP, http://www.fercap-sidcer.org/index.php), or the United States' Protection of Human Subjects ("Common Rule." Title 45 Code of Federal Regulations Part 46, https://www.hhs.gov/ohrp/regulations-and-policy/regulations/common-rule/index.html). In the national contexts where the three editors work (Denmark/Austria, Sweden, and USA), there are varying requirements for the review of human subject research procedures prior to the onset of any research activity. Such regulations are commonly set by governmental bodies. These guidelines safeguard those who are studied while making researchers reflect on not only who can be researched and in what ways but also what is good, fair and right to be researched. In particular, these guidelines apply when conducting research to obtain data through intervention or interaction with the individuals (e.g., use of a newly designed curriculum, teacher professional development) or collect identifiable private information (e.g., surveys).

Unsurprisingly, there are numerous books and chapters devoted to 'dealing with research ethics', because designing and conducting ethical research is crucial for a successful research study and, therefore, emerging researchers must learn about guidelines for research ethics and how to obtain an ethics review approval. However, such guidelines commonly constitute a regulated *minimum* of ethical consideration and may not necessarily consider the research methodologies used and the research topics dealt with in science education research. Regulations and ethical guidelines were traditionally developed from medical research ethics frameworks with the aim to mediate consideration of all risks to research subjects before the research is conducted (see the chapter prepared by Allison and Vogt 2020). This has had consequences in terms of the heavy emphasis on informed consent at the onset of data collection and participants' privacy in the pursuit of ethical consideration (Howe and Moses 1999). Science education research that is often qualitative and interpretive, and employs methodologies such as ethnography, case studies, interviews, or video based research that involve interactions with research participants are somewhat different from those methods used in medical research. The interactions with research participants in such methodologies range from talking with participants in a one-on-one interview setting to sustained interactions over longer periods of time aiming at the emancipation of teachers or students. The types of methodologies and types of interactions call for the development of a set of research ethical considerations that ensure researchers' responsibility and responsiveness within their research contexts and methodologies, which inevitably are different from those in medical research.

1.3   A Community Responsibility

Modern academic culture, which is shaped by a global neoliberal context, encourages (or even requires) many university-bound researchers to be productive in terms of publication rates (Luka et al. 2015). Productivity often equates with how fast and how many articles are published in peer-reviewed journals and how many research projects an individual has managed to secure external funding for. This academic culture appears to put individual researchers in a bubble that may obscure researchers' values of pursuing research with participants, various communities, and society at large in order to take collective responsibility as a research community. This means that we as a community of researchers should place more value on establishing and participating in joint conversations on how such conditions shape the ethical practices in science education research.

If we do not challenge our existing practices, we are likely to turn a collective blind eye on questionable practices. In a recent study the Norwegian ethical board (https://www.etikkom.no/en/news/news-archive/2019/40-percent-of-researchers-have-committed-a-qrp) published results from a survey showing that up to 40% of

Norwegian researchers have self-reported some forms of questionable research practices. Amongst issues reported by the researchers were that they have failed to inform stakeholders of their research projects about the limitations in the data analysis as well as having been influenced by the desires of funding bodies when designing their studies. In other words, there should be a critical need for science education researchers to engage in collective reflection on ethics and the challenges of acting with ethical responsibility and responsiveness.

As a community we need to think about modern day research realities that position us in a challenging context wherein, for instance, particular research topics or methods are preferred over others, and speed and number of publications are used as the sole measure of productivity. We need to consider how we can address such challenges and in what directions we should head. With this book we aim to engage in a conversation with the community of science education researchers so we can move from considering a mere compliance with governmental regulations as being ethical to collectively developing and sharing experiences and tools for reflection within the science education community. While the broad community of science education researchers may not all share the same ontological and epistemological assumptions, we believe that we, as a community, can and should focus on shared values and ethics and their implications for research practices despite such differences.

1.4 Methodological Reflections and the Need to Consider Ethical Implications

In science education research, theories and methodologies are continually evolving, which contributes to the emergence of new insights; this also creates tensions with respect to how research should be conducted. For example, in recent years there has been interest in theories such as new materialism (Milne and Scantlebury 2019) or actor-network theory and postphenomenology (Roehl 2012). The question is, whether theories that explore how students and teachers are affected (for example emotionally) by their interactions with materials require differentiated ways to research, especially since materials are seen here as actors that are put 'en par' with people. Another aspect is a push for collaborative relationships between researchers, science teachers, and students that seek to build relationships of trust. Participants in such kind or research are not seen as data sources and imply an entanglement of the hopes and wishes by both researchers and research participants. In this light it may be necessary to think about ontological questions concerning whether we should be referring to our participants as research *subjects* since even wording implies particular research assumptions and approaches.

Participatory methodologies and action research in science education can involve questions of authority and knowledge ownership and how we deal with relations between the involved actors. Since science education research often involves young people and their experiences, researchers, who are often positioned with authority and considered as more knowledgeable than their young research participants, are faced with challenges of seeking ways to include young participants' perspectives throughout all stages of research, from the formulation of research to dissemination of findings (Harcourt and Sargeant 2011). Furthermore, while it is important to ensure participants' anonymity and confidentiality of data, we want to pose questions on how to ethically work with participants, such as young learners or teachers who could become co-researchers and co-creators of what can be witnessed in the research settings, especially in participatory research.

1.4.1　Visual Data as an Example of How New Tools Create New Ethical Challenges

With the increase of more sophisticated data collection tools and analysis methods, the conditions for conducting ethical research have changed and call researchers to review again their accountability towards research participants (Levinson 2010). For instance, the emergence of the internet and the abundance of information that is made available (for example through blogs, social media, photos and videos, etc.) raise questions on participant recruitment practices and informed consent models, including participant expectations how they or those they representing may be benefiting from the proposed research. A particular interest is also the rise in visual data that is being collected to produce research that goes beyond the study of talk, that considers how teachers and their students interact with materials, display their emotions, or experience their learning environments, that are all factors that shape teaching and learning (Ritchie et al. 2013). This kind of research requires that researchers capture and study teachers and students' interactions in detail during the moments of teaching and learning. Facial expressions that give insight into how someone reacted to a given situation are difficult to share and discuss in text-only, traditional publication format. Sharing video data or images, however, means that people's identities may be revealed even if their names are not made public (for instance through the use of Facebook's algorithm *DeepFace* that allows for facial recognition). The speed at which facial recognition software develops suggests that in the near future such software may identify the identity of individuals at the click of a mouse. In this context, we may ask whether existing guidelines that are provided for research are keeping up with the modern realities of the visual presence of individuals in digital spaces.

1.5　The Particular Ethical Research Challenges for Science Education

All human endeavors involve values and the production and reproduction of values, and so does science education. The organized traditions of researching science teaching and learning have contributed to, and have been influenced by, particular methods, traditions and rituals. The philosophy and culture of science is shaped by 'logos' (Greek for the search for objectivity, facts and reason), and this is traditionally in conflict with the contentious nature of many areas and cultures of social science studies, including the cultures of science education research. A prominent example is the continuing discussion surrounding the conflict between science and religion. Contemporary issues that arise in discussions of climate change have much to do with the practices and the insights gained through science and our understanding about it. School science education has the possibility to contribute to dealing with potential conflicts i.e. by learning how to engage in informed discussions that may have the potential to identify conflict resolutions that are critical to the survival of our societies (Muralidhar 2019).

Science education has to deal with questions of epistemology and values since epistemological assumptions are a matter of ethical responsibility that afford and/or constrain our responsiveness to science issues. However, there is no 'correct' or 'absolute' way to look at the epistemological foundations of science education in reality and how this may shape our subjectivities. Subjectivities on that matter are important since they help us to ask whether science education as well as the research on its practices have to do with politics, neo-liberalism, sexuality or other categories that constitute our social order (Bazzul 2016).

We believe that attention needs to be paid to the kind of discourses and practices that produce certain ways of 'being', that does not exclude science education practices or the research around it. Limitations and affordances around our research practices that are delineated by discourses and the repetition of practices shape our perspectives of the phenomena we take an interest in. So, it seems important to spend more time on developing how researchers are constituted in and through their ethical research practices since this is still not discussed in-depth.

1.6 Outline of the Book

This book is organized into two parts: Part one is entitled *Challenging existing norms and practices* and part two *Epistemological considerations for ethical science education research*. Each part includes a number of contributions to the thematic focus and is rounded of by a reflection chapter where the authors departed from the points made in the previous chapters to present their own insights.

In *Challenging existing norms and practices* the discussions of the contributing authors are focusing on questions like: What are the conditions of knowledge that shape ethical decision making? Where is this kind of knowledge coming from? How is this knowledge structured, and where are the limitations? How can we justify our beliefs concerning our ethical research actions? As well as the issues that have to do with the creation and dissemination of knowledge through research approaches in science education. In reflecting upon methodological considerations fundamentally philosophical questions of the relevance of research ethics are raised by Antje Gimmler in chapter 2 (Gimmler 2020). Questions are also raised concerning the range and varieties of methodological practices of research in science education from historically oriented science education research in chapter 3 by Allison and Vogt (2020), to ethnographies of education in chapter 4 by Minjung Ryu (2020) and in particular the values and knowledge at stake in researching educational practices in chapter 5 by Johansen and Anker (2020), how these values and norms are entangled with the science content and how this becomes visible when dealing with contentious contents such as in sexuality education in chapter 6 by Orlander and Lundegård (2020). The first part of the book concludes with the commentary chapter 7 by Jaume Ametller (2020), who, prompted by the previous chapters, reflects on how to engage with the political and onto-epistemological ideas related to the ethical challenges we face in science education research.

In *Epistomological considerations for ethical science education research* the discussions of the contributing authors are centered around the norms and practices of conducting science education research in regard to methods, validity and scope. In chapter 8, Andrée et al. (2020) examine the symmetry of relations in science education research contrasting ontological with epistemological and methodological values to reflect on research practices. Adams and Siry (2020) examine in chapter 9 the Athenticity Criteria first described by Guba and Lincoln (1989) and reflect on how this supporting their science education research to be transformative and authentic. Scantlebury and Milne (2020) explain in chapter 10 what the ethical consequences the adoption of a post-humanist approach after Karen Barad mean. The chapter describes that this theoretical approach identifies human action as being emergent to allow researchers to identify material-discursive practices. In chapter 11 Jaakko Hilppö and Stevens (2020) zoom in on science education research that utilises video recording to allow for the capture of students' voices to make them agents of their own practices. Focusing on material ethics Kathrin Otrel-Cass (2020) argues that the practice of conducting research ethically is an ongoing practice that is difficult to imagine in its full spectrum a priori, but requires ongoing reflections and communications between researchers and their participants. The second commentary chapter that concludes the book is written by Martin Riopel (2020) and acknowledges a shift of focus in the chapters from macro-level considerations to micro-level considerations. In this closing chapter, Riopel argues that this can be interpreted primarily as a sign of maturity in the field but also as an alignment with some of the challenges of the current society.

References

Adams, J., & Siry, C. (2020). Living authenticity in science education research. In K. Otrel-Cass, M. Andrée, & M. Ryu (Eds.), *Examining research ethics in contemporary science education research*. New York: Springer Publishing Company.

Allison, J., & Vogt, M. (2020). Reflections on research ethics in historically oriented science education research in Canada. In K. Otrel-Cass, M. Andrée, & M. Ryu (Eds.), *Examining research ethics in contemporary science education research*. New York: Springer Publishing Company.

American Association for the Advancement of Science. (1993). *Benchmarks for science literacy*. New York: Oxford University Press.

Ametller, J. (2020). Challenging existing norms and practices: Ethical thinking at the science education research boundaries. In K. Otrel-Cass, M. Andrée, & M. Ryu (Eds.), *Examining research ethics in contemporary science education research*. New York: Springer Publishing Company.

Andrée, M., Danckwardt-Lillieström, K., & Wiblom, J. (2020). Ethical challenges of symmetry in participatory science education research – Proposing a heuristic for ethical reflection. In K. Otrel-Cass, M. Andrée, & M. Ryu (Eds.), *Examining research ethics in contemporary science education research*. New York: Springer Publishing Company.

Bazzul, J. (2016). *Ethics and science education: How subjectivity matters*. Cham: Springer International Publishing.

Economic and Social Research Council. (2015). *ESRC framework for research ethics: Updated January 2015*. Retrieved on 11/23/2019 from https://esrc.ukri.org/files/funding/guidance-for-applicants/esrc-framework-for-research-ethics-2015/

Gimmler, A. (2020). The relevance of relevance for research ethics. In K. Otrel-Cass, M. Andrée, & M. Ryu (Eds.), *Examining research ethics in contemporary science education research*. New York: Springer Publishing Company.

Guba, E. G., & Lincoln, Y. S. (1989). *Fourth generation evaluation*. London: Sage publications.

Harcourt, D., & Sargeant, J. (2011). The challenges of conducting ethical research with children. *Education Inquiry, 2*(3), 421–436.

Hayles, N. K. (2017). *Unthought: The power of the cognitive nonconscious*. Chicago: University of Chicago Press.

Hilppö, J., & Stevens, R. (2020). Students' ethical agency in video research. In K. Otrel-Cass, M. Andrée, & M. Ryu (Eds.), *Examining research ethics in contemporary science education research*. New York: Springer Publishing Company.

Howe, K. R., & Moses, M. S. (1999). Chapter 2: Ethics in educational research. *Review of Research in Education, 24*(1), 21–59.

Johansen, G., & Anker, T. (2020). Science education practices: Analysing values and knowledge when conducting educational research. In K. Otrel-Cass, M. Andrée, & M. Ryu (Eds.), *Examining research ethics in contemporary science education research*. New York: Springer Publishing Company.

Leathwood, C., & Read, B. (2013). Research policy and academic performativity: Compliance, contestation and complicity. *Studies in Higher Education, 38*(8), 1162–1174.

Levinson, M. P. (2010). Accountability to research participants: Unresolved dilemmas and unravelling ethics. *Ethnography and Education, 5*(2), 193–207.

Luka, M. E., Harvey, A., Hogan, M., Shepherd, T., & Zeffiro, A. (2015). Scholarship as cultural production in the neoliberal university: Working within and against 'deliverables'. *Studies in Social Justice, 9*(2), 176–196.

Millar, R., & Osborne, J. F. (1998). *Beyond 2000: Science education for the future*. London: King's College. Retrieved on 11/21/2019 from https://www.nuffieldfoundation.org/sites/default/files/Beyond%202000.pdf.

Milne, C., & Scantlebury, K. (2019). *Material practice and materiality: Too long ignored in science education*. Cham: Springer.

Muralidhar, K. (2019). Introduction. In K. Muralidhar, A. Ghosh, & A. K. Singhvi (Eds.), *Ethics in science education, research and governance*. New Delhi: Indian National Science Academy.

NESH (The National Committee for Research Ethics in the Social Sciences and the Humanities, Law and Theology). (2016). *Forskningsetiske retningslinjer* [Guidelines for research ethics]. Oslo: The Norwegian National Research Ethics Committees.

Orlander, A. A., & Lundegård, I. (2020). Sex education — Normativity and ethical considerations through three lenses. In K. Otrel-Cass, M. Andrée, & M. Ryu (Eds.), *Examining research ethics in contemporary science education research*. New York: Springer Publishing Company.

Osborne, J., & Dillon, J. (2008). *Science education in Europe: Critical reflections*. London: Nuffield Foundation. Retrieved on 11/23/2019 from https://www.nuffieldfoundation.org/sites/default/files/Sci_Ed_in_Europe_Report_Final.pdf.

Otrel-Cass, K. (2020). The performativity of ethics in visual science education research: Using a material ethics approach. In K. Otrel-Cass, M. Andrée, & M. Ryu (Eds.), *Examining research ethics in contemporary science education research*. New York: Springer Publishing Company.

Pereira, Â. G., Vesnić-Alujević, L., & Ghezzi, A. (2014). The ethics of forgetting and remembering in the digital world through the eye of the media. In *The ethics of memory in a digital age* (pp. 9–27). London: Palgrave Macmillan.

Riopel, M. (2020). Methodological ethics considerations in science education research: Symmetric, authentic, material, adaptive and multidisciplinary. In K. Otrel-Cass, M. Andrée, & M. Ryu (Eds.), *Examining research ethics in contemporary science education research*. New York: Springer Publishing Company.

Ritchie, S. M., Tobin, K., Sandhu, M., Sandhu, S., Henderson, S., & Roth, W. M. (2013). Emotional arousal of beginning physics teachers during extended experimental investigations. *Journal of Research in Science Teaching, 50*(2), 137–161.

Roehl, T. (2012). From witnessing to recording–material objects and the epistemic configuration of science classes. *Pedagogy, culture & society, 20*(1), 49–70.

Ryu, M.-J. (2020). Ethical considerations in ethnographies of science education: Toward humanizing science education research. In K. Otrel-Cass, M. Andrée, & M. Ryu (Eds.), *Examining research ethics in contemporary science education research*. New York: Springer Publishing Company.

Scantlebury, K., & Milne, C. (2020). Beyond dichotomies/binaries: 21st century post humanities ethics for science education using a Baradian perspective. In K. Otrel-Cass, M. Andrée, & M. Ryu (Eds.), *Examining research ethics in contemporary science education research*. New York: Springer Publishing Company.

Schreiner, C., & Sjøberg, S. (2004). *Sowing the seeds of ROSE. Background, rationale, questionnaire development and data collection for ROSE (The Relevance of Science Education) – A comparative study of students' views of science and science education. Acta Didactica 4/2004.* Oslo: University of Oslo.

The Norwegian Ethics Research Committees. (2019). Retrieved on the 1.11.2019 from https://www.etikkom.no/en/news/news-archive/2019/40-percent-of-researchers-have-committed-a-qrp/

Kathrin Otrel-Cass, is a Professor of education and digital transformations at the University of Graz, Austria. Her research interests are often of interdisciplinary in nature with focus on digital visual anthropology and variety of qualitative, ethnographic methodologies. She works with various practitioners and experts in environments where people are working with science/technology/engineering practices or their knowledge products. Her research is often set in schools but is not exclusive to those environments. Her research interest in visual ethnography has led to the establishment of a video research laboratories at Aalborg University and the University of Graz with a focus on the organized analysis of video recorded data and ethical research practices involving visual data. Kathrin is also a member of the Human Factor in Digital Transformation research network at the University of Graz.

Maria Andrée is an Associate Professor of Science Education at the Department of Mathematics and Science Education at Stockholm University, Sweden. Drawing primarily on socio-cultural theory, her research focuses on science education practices and the conditions for students' participation and learning, particularly in relation to questions of science curriculum, scientific literacy and citizenship. She primarily works with ethnographic and design-based research studies in science education. She has also pursued a line of research concerning the involvement of external actors in science, technology and mathematics education. She is currently one of the scientific leader of *Stockholm Teaching & Learning Studies* – a platform for research in collaboration with teachers – designed to initiate, support and conduct small-scale classroom-based didactic research on teaching and learning. She has published in peer-reviewed journals including *International Journal of Science Education, Research in Science Education, Cultural Studies of Science Education*, and *Journal of Curriculum Studies* among others.

Minjung Ryu is an assistant professor in Chemistry and Learning Sciences at University of Illinois at Chicago, USA. Her research focuses on STEM learning and participation of cultural and linguistic minority students. Employing ethnography and discourse analysis, she examines how racial, ethnic, and linguistic minority students engage in STEM discourses using multilingual and multimodal means and what are ways to design learning environments to improve these students' learning experiences. Within this research interest, she has worked with resettled Burmese refugee teens in a community-based afterschool program in USA where the teens learn STEM knowledge to transform their communities and global societies. She also has collaborated with high school science teachers to develop instructional materials and practices to support English learners in linguistically superdiverse classrooms. Minjung has published in *Journal of Research in Science Teaching, Science Education*, and *International Journal of Science Education*.

Part I
Challenging Existing Norms and Practices

Chapter 2
The Relevance of Relevance for Research Ethics

Antje Gimmler

2.1 Introduction

Relevance of research does not usually count as a central criterion of research ethics. The question of relevance of research often has its place either in relation to scientific criteria of progress and accumulation of knowledge or within a discussion of the relation of research to the needs of society. In the latter kind of discussion, relevance equals all too often merely usefulness, sometimes understood in a crude economic sense. Thus, the question of relevance sits within a very mixed field: between inner academic standards for research, and the public, political or economic demands that contribute to pressing problems or economic growth. However, relevance is a relevant issue for research, and I will argue that a clear notion of the relevance of research could help researchers to navigate this tension and to link inner academic standards with social responsibility.

As aforementioned, research ethics often does not entail considerations on the relevance of research. Research ethics as such is typically limited to topics like harm to participants, informed consent, privacy issues, deception, and fraudulent research as well as harm to the environment (Williams 2016, p. 42). Good examples are the very detailed guidelines from the *Office of Research Integrity*[1] at the University of Pittsburgh or the *European Textbook on Ethics in Research*,[2] published by the European Commission (EU). These issues and the guidelines for research ethics and good scientific practice are definitely most relevant for the

---

[1] https://www.orp.pitt.edu/rcco-offices/research-integrity

[2] https://ec.europa.eu/research/science-society/document_library/pdf_06/textbook-on-ethics-report_en.pdf

---

A. Gimmler (✉)
Aalborg University, Aalborg, Denmark
e-mail: gimmler@hum.aau.dk

© Springer Nature Switzerland AG 2020
K. Otrel-Cass et al. (eds.), *Examining Ethics in Contemporary Science Education Research*, Cultural Studies of Science Education 20,
https://doi.org/10.1007/978-3-030-50921-7_2

conduct of research, both for researchers and students who learn to evaluate and perform research. As Janet A. Kourany (2011) argues, it is important to implement these guidelines and to acknowledge the normativity of this project, which is "goals and responsibilities scientists ought to set for themselves." (Kourany 2011, p. 378). However, the use of guidelines is not without problems, and whether ethics committees are always able to evaluate the methodological and practical subtleties of a research project can be doubted (Hammersley 2009). Taking this difficulty into account, I will argue that the scope of what research ethics comprises of should not be conceptualized to be unnecessarily narrow (see also Williams on this point, Williams 2016, p. 42 f.). The question of relevance should be seen as part of a broader understanding of research ethics that for instance also includes social responsibility. There are many problems connected to this broader understanding of research ethics and here I only want to highlight one of them: relevance could be, and has been, understood as purely instrumental to direct economic and political interests and thereby collide with the freedom of research, a value that is a necessary presupposition for research. From my point of view this reduction to economic and political interests is a misconception of what relevance really means for research. Reduced to usability, this type of economic and politically restricted notion of relevance tends to undermine the dynamics of research as such.

Researchers in science education research are not an exception to these tensions that are part of public debates and policies. Firstly, they not only use the ethical guidelines for good scientific practice, as other research professions, they are also socialized to follow professional objectives and to adopt a professional view on what role research and science plays and should play in society. Over time they adopt a professional identity that makes it difficult to see e.g. school policies that have been not intended by their research (Mitcham 2003). Secondly, politics also influences science education researchers, and especially in the last 20 years politicians and economists ask for more students in STEM subjects (in German speaking countries called MINT subjects). It has become a highly political question how to interest pupils and students for the natural sciences and technology. Cases in point are the initiative *National Natural Science Strategy*[3] that the Danish Ministry of Education launched in 2018, the *Independent Review of Primary Curriculum*[4] (2016) in the UK that informs school policies and the EU report *Science Education for responsible Citizenship*[5] (2015). Each of these reports highlights the role of science education for addressing future societal challenges. They recommend strengthening numeracy and ICT competences of pupils and students with the purpose to meet future demands of industry and to enable innovation. Albeit these reports focus on STEM subjects it is also true that there is at least some emphasis on the arts, on creativity, and on social competences; subjects which supposedly contribute positively to innovative societies. Looking at these statements and initiatives for

---

[3] https://dea.nu/sites/dea.nu/files/sammenfatning_stem_web_engelsk_2.pdf

[4] https://dera.ioe.ac.uk/30098/2/2009-IRPC-final-report_Redacted.pdf

[5] http://ec.europa.eu/research/swafs/pdf/pub_science_education/KI-NA-26-893-EN-N.pdf

science education that are the result of political decisions, one can diagnose a certain political pressure that might have an impact onto the theoretical and practical framework of how to teach and research science education. In this political climate the question of relevance of science education research is not only a problem of inner academic contributions to scientific progress but also a question of broader ethics and social responsibility.

Usefulness and purpose of research and its relation to society is a highly debated issue in science studies and in philosophy (Resnik 2007; Carrier and Nordmann 2011). One of the most distinguished positions concerning the problems related to the relevance of research is the pragmatic understanding of research as part of a collective achievement (Bohman 1999). Pragmatists hold the claim that relevance of research and performing sound research in terms of scientific methodology go hand in hand. The pragmatists' focus on relevance does not render ethical guidelines superfluous, but ethical guidelines should be seen as part of a broader picture. In this article I shall argue that philosophical pragmatism offers researchers an understanding of relevance that is neither reduced to the inner scientific standards of progression nor to the external demand for economic value. Pragmatism is a philosophical tradition that has its roots in the nineteenth century with the work of Charles S. Peirce, William James and John Dewey and had been influential during the first third of the twentieth century. Pragmatism had then been replaced by analytical philosophy, though exhibited still a certain 'underground' influence. In pedagogy and sociology, pragmatism has never stopped to be a source of inspiration and Dewey's democratic and experience-based pedagogy had been most influential. Since the 1980s philosophical pragmatism has experienced a revival. It was, among others, the neo-pragmatist Richard Rorty and the pragmatic language philosopher Robert Brandom who revitalized pragmatic thinking. As a result, pragmatic philosophy today is a vital philosophical movement with a broad spectrum of different positions (Misak 2013). Relevance of research and science is a topic that is central to pragmatists. For Dewey, relevance of research is a most important issue if we expect that research successfully contributes to the common good. However, how do we decide which research question is relevant and which is not? Are there any criteria?

In what follows, I shall take up the question of relevance of research firstly by giving a sketch of how relevance is treated in common textbooks of research design. I shall then use this analysis as the background for the second part, where I shall introduce the pragmatic maxim of Charles S. Peirce as an alternative way of elucidating how relevance plays a role in defining the scope of research and how this connects to the perception of the research topic as such. In the third part I shall then utilize the philosophy of John Dewey to continue Peirce's approach by arguing that relevance indeed is a relevant topic for education science research ethics, and that especially in science education research the interdependency of science and society should not be downplayed by simply demanding so-called social impact of science without being aware what social impact actually means.

2.2 Relevance Between Inner-Academic Goals and the Needs of Society

As aforementioned, relevance of research is usually not a topic of research ethic guidelines. This picture changes when we look at introductions to research design and research as such. They often contain a chapter on research ethics and on the relevance of research as well. In what follows I shall refine myself to introductions to research design for the social sciences because the vast majority of science education research makes use of methods and methodology stemming from the social sciences. This selection claims neither to be representative nor exhaustive, but shall give a first understanding of the topic at stake. In these books on research design relevance is connected to the choice of research design, and what counts as relevant research is dependent upon the methodology and philosophy of science positions adopted by the researcher. If one holds a philosophy of science position that conceptualizes science as puzzle-solving and testing of theories, like Karl Popper does, then research is relevant if it contributes to solve the puzzles that arises from theories. Pragmatists on the contrary would always think of relevance as something that is deeply connected with 'real-life-problems' and would evaluate science as puzzle-solving by its contribution to solving real-life problems. Other philosophers have highlighted also that ethics and philosophy of science are not always neatly separable (Denscombe 2002; Tuana 2013) and in the next section we will see how the pragmatists tie this knot by seeing ethics, social responsibility and sound research interconnected.

I take my starting point with the useful book on *Ground rules for good research* by Denscombe who includes the question of relevance of research. From his perspective, relevance is part of the research design and those background assumptions about the nature of research stemming from philosophy of science. Generally, Denscombe says, "the users of the research stand to gain much by being presented with clear statements about the relevance of the research" (Denscombe 2002, 44). The statements needed would then tell the reader/user of the research about values, sources and 'vision' of the research undertaken. Denscombe names four types of relevance (see p. 45–49): relevance to existing knowledge, relevance to a practical need, relevance in terms of the timeliness of research, relevance to the researcher.

One candidate for the criterion of relevance that Denscombe names, is relevance to a researcher. Here, we are confronted with the arbitrariness connected to the more or less subjective reasons why and how a researcher addresses a specific research question. As also Maurice Punch (Punch 1994) highlights, there are circumstances like geographic proximity and possible access, e.g. to a school or a group of students, that might determine the research object. Although subjective motivation and arbitrary circumstances influence the choice of the research topic and object under investigation, the form of relevance stemming from this kind of reason for undertaking specific research are by no means good reasons for justifying important choices made in the research process. These motivations do not fit to the general goal of science to produce objective knowledge, if objectivity means to "eliminate personal, social, economic, and political biases from experimental design, testing, data analysis and interpretation, peer review, and publication" (Resnik 2007, p. 46). It is also right that researchers often chose what fits into their research agenda, what boosts their careers, or what gives them more prestige, however, these motivations cannot count as proper criteria for the relevance of research; neither ethically nor if measured by purely inner academic standards is relevance to the researcher a valid justification for the relevance of research.

The next type of relevance, timeliness, is difficult to grasp. In Denscombe's definition "matters that are high on the agenda of current concerns" (Denscombe 2002, p. 47) make research timely. He also names that one should avoid research being "overtaken by events" (Denscombe 2002, p. 48). Besides clear-cut cases like investigating into a school policy that has been abolished (supposed the research is not historically oriented), it is difficult to say if research is timely or not. Successful research funding might be an indicator for timeliness, however, research funding institutions can also be utterly conservative and not acknowledging research that presents new views and might be 'timely'. It is difficult to define in each case whether the research is timely and it is an open question as to who decides which research is timely. In the case that the justification for relevance is relegated to the public, then to call research timely seems to be just another way of saying that relevant research meets the practical needs of a society, which is another type of relevance Denscombe discusses.

The next type Denscombe names, is relevance of research in relation to existing knowledge and this is traditionally one of the main criteria used to define the relevance of research. This type clearly refers to the accumulation model of science and to what has been coined 'puzzle solving'. Research in this sense has the aim of filling the missing gaps in existing research, and relevance becomes a matter of how new knowledge contributes to and refines existing knowledge. Hammersley (1995) has called this the 'disciplinary model' because application of research is neither directly nor immediately intended. This is a more conservative way of perceiving research where science rests on a "building block approach" (Denscombe 2002, p. 46) and new knowledge is always an answer to a question posed within the vocabulary of already established knowledge. This type of research relevance claims to be neutral to socio-political circumstances and demands. However, this classical understanding of science and research of being neutral to values, for instance research that relies on positivistic premises, has been the target of much criticism. The claim of positivism (as well as critical rationalism) that appropriate methods are able to eliminate values has been problematized by philosophers of science such as Thomas Kuhn or Joseph Rouse. From the view point of the social sciences the positivistic ideal of science cannot be transferred to social science research. G. and J. Payne (2004) highlight that the "choice of topic, the theories brought to bear, how research questions are posed, kinds of data collection and analysis, and the construction of conclusions, are all stages where values can and do intervene." (Payne and Payne 2004, p. 154) While Payne and Payne refer to values that influence research from within, others like Clifford Christians go even further and make the claim that research often lacks a "research ethics in which human action and conceptions of the good are interactive" (Christians 2005, p. 158), thereby applying an understanding of research that follows external values and thereby exhibiting a particular social responsibility. Christians refers to research in the name of a feminist agenda as a role model for this type of intervening research that demonstrates a clear value orientation. He advocates relevance of research in the name of a good cause or a political ideal. One could also point to action research as a research approach that has the explicit goal to include schools, pupils, teachers and parents in an intervening and transformative research process.

If values are indeed a necessary part of research, relevance cannot not be defined with reference to purely inner academic puzzle solving and filling the gaps of the edifice of science. Skepticism about the pure model of science advocated by positivists is widespread. Also Stuart Farthing (2016) in his introduction to research design in urban planning discusses the role of values in research. He refers to Max Weber and his nuanced understanding of values in research as well as his notion of value relevance (Farthing 2016, p. 183). Weber has argued that a distinction is needed between political values and 'Weltanschauungen' (Weber 1988, p. 153) that might influence research (and its results) on the one hand, and those values (Wertideen) that are necessary in order to choose the topic of research and methods on the other. While the first leads to deeply biased research, the latter function of values is unavoidable and part of an overt process of checking one's 'idealtypes' (Weber 1988, p. 199) against empirical facts. Weber describes this process vividly as one of several iterations where the guiding values (Wertideen) become transparent and open for correction. From Weber's point of view there are clearly limits of how much researchers ought to be involved – even in the name of the good deed and high moral values – in their research objects. Weber would accuse Christians and parts of action research to confuse the guiding research values with direct political and social partialities. For Weber research is not value-free with respect to the criteria for selection and construction of research ideas and topics, but should be free from the direct influence of political or economic interests that impose certain research topics and dictate the direct application of political ideals, religious beliefs, or philosophical 'Weltanschauung'. That values in research cannot be avoided, is also mentioned by Alan Bryman in his much used handbook on *Social Research Methods* (Bryman 2008, p. 24 f.). Values are part of research in form of the formulation of the research design, the data collection methods, the analysis of data etc. Thus, value-free research is, even for those who like Weber do not advocate research as a form of political action, impossible. To sum up: The disciplinary model of science and its understanding of relevance is problematic because this model of science is thought to be value neutral. Here, relevance relies purely on the contribution of science and research to puzzle solving and in the end on the presupposition that scientific methods uncover the truth of reality. Among others, Weber has outlined that value neutralness as such – and this would also be applicable to science education research – is not possible. Value neutralness would bereave knowledge for what it is worthwhile, namely its relevance. The distinction between external and internal values Weber introduces, is here helpful. However, how to distinguish these two value orientations and to keep them apart in the research process remains an open question.

The last type of relevance Denscombe names is the relevance that meets practical needs (Denscombe 2002, p. 46); an approach that sounds promising, and at first glance seems to be a good candidate for a meaningful criterion for relevance of research. As Farthing states "the moral or ethical argument is that researchers ought to help practitioners make practice more effective. (…) research should be 'relevant' to their needs and should provide evidence to underpin policy and practice" (Farthing 2016, p. 182). However, the complex relation between society and research implies many questions: Who articulates the needs of a society — politicians, administration, media, or citizens? And what are practical needs? Are there other needs of a non-practical kind? If research is not meeting the needs of a society, does this render the research irrelevant? Is the researcher's role reduced to deliver solutions for ready-made problems? Denscombe emphasizes that "the significance of the problem needs to be established" (Denscombe 2002, p. 47) and this indicates that the relevance of research still is dependent upon the formulation of the problem by the researcher. Unfortunately, the relevance criterion of practical needs leaves us with a circular argumentation: the practical needs that are 'out there' in society have to be evaluated and established as research problems by researchers. Then, it is again the researcher who applies the criterion of relevance to what she identifies as possible practical needs. This is most unfortunate for the claim that practical needs provide a viable criterion for relevance.

Another problem of relevance of research as answering practical needs is related to the responsibility researchers might then have for their research. Responsibility for research would be a logical consequence if research should answer to the practical needs of a society. However, how can researchers take responsibility if the effects and consequences of research are not known and often take us by surprise. If there is a principal uncertainty about the consequences of research, are researchers then responsible for these unknown and unintended consequences? From a positivistic point of view, as has been already discussed, values and ethical responsibility should not be imposed on researchers who are doing basic and pure academic research or, as it has been called today, science of the modus 1 type (Nowotny et al. 2001). However, this rigid positivistic position has met critique. And especially on the background of a fundamental change in the way research and science takes place today it is highly questionable whether the clear distinction between values and facts in a strict positivistic sense can be upheld. Basic research which pursues a scientific investigation in order to enciphers nature's book (or human nature) is not neatly separated anymore from applied research and possibly never has been. Thus, the question of responsibility and values of research appears in a different light (Nowotny et al. 2001, p. 10 ff.). If Nowotny et al. are right and many studies support their diagnosis, then research today cannot just hide behind the allegedly value-neutral basic research and leave the question of responsibility to practitioners and society itself. As the philosopher of science Martin Hollis puts it: "Expertise carries special ethical responsibilities" (Hollis 1994, p. 205) It is well known that the Nuremberg Code (1947) as a reaction to the atrocities in the name of research committed by the Nazi Regime was the starting point for an advanced ethical awareness of researchers and scientists. The same goes mutatis mutandis for the

Russell-Einstein Manifesto in 1955 which was an reaction of scholars and scientists to the threats of nuclear war. As a result it became clear that ethical issues and responsibility for research is not something that can be outsourced to the public and politicians.

While there are many pitfalls connected to this idea that relevant research needs to meet the practical needs of society there is of course also a critical gist to it. Farthing (2016) presents the critical model of relevance as a form of engaged research that addresses so-called practical needs not directly, but is committed to uncover underlying power relations and thereby elucidating societal injustice or problems in general. Furthermore, the idea of relevance as meeting practical needs gains some of its persuasive power from the fact that the gap between academic research and practical application is bridged by research for and with society. This is also what Bent Flyvbjerg recommends in order to make "social science matter" (Flyvbjerg 2001): "We may transform social science to an activity done in public for the public, sometimes to clarify, sometimes to intervene, sometimes to generate new perspectives, and always serve as eyes and ears in our ongoing efforts at understanding the present and deliberating about the future. We may, in short, arrive at a social science that matters." (Flyvbjerg 2001, p. 166). Social science matters if it is able to contribute fruitfully to public deliberations, not merely about professional matters but as well about the values that are at stake for a society. The Aristotelian phronesis is Flyvbjerg's tool to safeguard context-sensitive research that still has enough explanatory power to produce knowledge that informs and enlightens policies, planning, and society. This view is applicable to science education research that often deals with questions not only of academic but public interest, such as the introduction of digital learning tools for teaching.

In this short analysis of the role of relevance to research ethics it should become clear that the topic of relevance of research is situated in a very complex field. In one end of this field, we find the more narrow ethics of research in form of guidelines and checklists. In the other end, we find much broader considerations about the responsibility of research and related consequences, often combined with epistemological and methodological considerations about objectivity, values and research design in general. Critical research and research that meets the needs of a society are possible candidates for an ethically substantial criterion of relevance. However, the tension between external values stemming from society and internal values that govern the scientific process has not been solved. Nancy Tuana (2013) suggests three main dimensions of how ethics is present in research: procedural ethics deals with the defined goals of responsible conduct of research (RCR), intrinsic ethics with issues that are internal to the research design, and extrinsic ethics which refers to the impact of research on society as well as the demands of society (Tuana 2013, p. 1961ff). Applying her terminology on the short analysis of the different understandings of relevance of research it can be stated that the problem of relevance arises at the intersection of the intrinsic and extrinsic dimension of ethics in research, thereby blurring the lines between societal demands, ethical norms and epistemological issues. For those advocating socially responsible research that is problem oriented and transcends traditional disciplinary boundaries by suggesting a broader interdisciplinary set-up for research this is not a bad thing. Actually, interdisciplinarity and epistemological as well as ethical-social awareness is seen as a way to put science in a fitter state in order to answer the pressing and wicked problems the world is facing (Frodeman et al. 2001). In the following two parts I shall show how the pragmatists contribute to this newer approach to interdisciplinarity and ethical orientation of research by applying the concept of relevance.

2.3 The Pragmatic Maxim of Peirce as a Test for Relevance

In his article "How to Make Our Ideas Clear" (Peirce 1935), Charles Sanders Peirce introduced the pragmatic maxim as a methodological tool to elucidate the meaning of concepts and thereby circumvent skeptical problems related to traditional epistemology and the theory of truth. This famous article belongs to Peirce's writings on pragmatism, and most Peirce scholars argue that Peirce's pragmatism is part of his much broader system of semiotics (e.g Pietarinem 2005; Haack 2018). However, focusing on the question of relevance of research I shall confine myself to Peirce's pragmatic maxim without taking his semiotics or the vast amount of literature that deals with the logical and epistemological dimensions of Peirce's maxim into account. What I suggest here is to use the pragmatic maxim as a principle of relevance for science education research. Peirce wants us to understand that we only know the meaning of a concept if we know what practical consequences we can expect if the hypothesis derived from the conception is tested in reality. This test in reality leads to what Peirce calls the third grade of clarity. In terms of relevance, only those subjects are fruitful research subjects that can lead to hypotheses that have possible practical consequences: "If a belief has no consequences – if there is nothing we would expect would be different if I were true or false – then it is empty or useless for inquiry and deliberation" (Misak 2013, p. 30). As has been highlighted by several Peirce scholars, the pragmatic maxim is not only a tool to know the meaning of a concept, but also to identify meaningless concepts. In Peirce's own words "to show that almost every proposition of ontological metaphysics is either meaningless gibberish (…) or else downright absurd." (Peirce 1935, p. 423) In this sense, the pragmatic maxim is a research guiding principle that helps researchers to identify the scope of relevance of their intended research. This is not so farfetched taking into account that Peirce and the other pragmatist saw the necessity to renew philosophy, the social science as well as humanities and thereby make them matter.

Peirce shares this critical attitude towards classical philosophy with the other pragmatists and he was convinced that a renewed philosophy could inform the sciences and have a great and useful impact on society. This program to innovate philosophy appears most explicitly in John Dewey's writings. A point in case is Dewey's book on "Reconstruction in Philosophy" (Dewey 2008a) where he recommends that philosophy should stop metaphysical and apriori thinking and instead adopt a more empirically saturated stance. Following Dewey, inquiry and the process of knowledge acquisition will appear in a different light, now taking knowledge as a part of practices, although an important and necessary part. When practices are the start and endpoint of inquiry, the question if either inner academic or external values establish the research topic becomes superfluous. Knowledge and theories are means to realize better practices and are not ends in themselves. As Dewey points out it is important "to establish a criterion which would enable one to determine whether a given philosophical question has an authentic and vital meaning, or on the contrary, it is trivial and purely verbal; and in the former case, what interests are at stake, when one accepts and affirms one or the other of the two theses in dispute." (Dewey 1988, p. 8) Peirce thought of his pragmatic maxim as such a principle.

As a pragmatic strategy Peirce recommends to replace apriori "ontological metaphysics" (Peirce 1935, p. 423) with an understanding of concepts that are not representations but foremost prescriptions for the ways the objects of our concepts would act within the realm of experience:

> *Consider what effects, that might conceivably have practical bearings, we conceive the object of our conception to have. Then, our conception of these effects is the whole of our conception of the object* (Peirce 1935, p. 402).

This is a complicated sentence and I shall first untangle the different parts with the help of the example given by Peirce himself: we have an object, a diamond, and we would like to know what kind of object this is. We should not start with apriori reasoning about the meaning of a diamond; instead we should start to conceptualize the object in a way that enables us to make propositions that lead to consequences. Instead of only reflecting about these propositions as such (with deductive logic e.g.), we should consider them as prescriptions for action which tell us about the effects that the object of our concept has: effects that 'conceivably have practical bearings'. We are looking for a conception of the object that not only has effects but also has effects that have practical bearings. It is impossible to know the possible practical consequences of the effects an object exhibits beforehand; therefore we have to develop an action-guiding hypothesis, where we investigate what would happen if a certain hypothesis is right or wrong. Note that we are not interested in all possible effects an object can have, but in those effects that have practical consequences. Then, if we have found out which effects these are, 'our conception of these effects is the whole of our conception of the object'. The concept is dependent upon the way we frame the inquiry.

In the case of the diamond, we start by formulating a concept about the object by putting possible propositions in the form of different hypotheses forward that could allow getting to know effects of the object. In this example, Peirce chooses hardness and we can test for these effects. These effects have practical bearings. They make a difference to how diamonds act in relation to other material. The diamond discriminates from other objects, e.g. a stone made of glass. While the glass stone will not be able to scratch other materials, the diamond will turn out to be able to scratch all other materials (including the glass stone) and we can now claim that the diamond is the hardest material (of all the tested materials). Hardness is a characteristic and an effect of this material that has practical bearings. While Peirce used the pragmatic maxim to clarify the meaning of a concept, we use the pragmatic maxim to clarify what kind of research is relevant and which is not. The preliminary answer is: there are good reasons to call research that consists of hypotheses that look for effects with practical bearings relevant research.

There are several implications related to this use of the pragmatic maxim to clarify the question of relevance of research. First of all, going back to the example, we have to consider that there are many other characteristics that also could form true propositions and hypothesis about the diamond and would have some sort of effects. A diamond glitters, is cold, or reflects light in a certain way. In different contexts these characteristics might be relevant. However, research is reductive in the sense that only one hypothesis after the other can be investigated. Which hypothesis to choose is a serious question. Peirce has contributed to this problem with the logic of abduction, a form for reasoning that is also called the inference to the best explanation. He has also formulated an economic principle for the scientific research process (Rescher 1976) that simply states that some inductive procedures are not worthwhile to perform, either because the outcome would make no significant and relevant difference or too many inductive tests would be afforded (Rescher 1976). To choose a hypothesis wisely is not only important in order to get a result, but in order to get a relevant result. Hypothesizing possible effects that have practical bearings guides us to a relevant problem that is worth conceptualizing as a research question.

Secondly, the research situation we find ourselves in has to be one where the practical bearings some effects may have, are unknown in the first place. The reason why we undertake research is a lack of knowledge about the outcome of possible actions connected with the object under investigation. The situation has to be an "indeterminate situation" (Dewey 1986, p. 109), in the words of John Dewey. The indeterminate situation does not present the research problem unmediated or directly, rather it is the researcher's task to define the problem stemming from the indeterminate situation. A scientifically fruitful indeterminate situation is different from the futile and artificial doubt Peirce accuses Descartes of. Real doubt "prompts real inquiry" (Haack 2018, p. 214). The starting point with an indeterminate situation has the advantage of leading to knowledge that should solve the indeterminate situation. The indeterminate situation is open for many different problem definitions and thus different guiding hypothesis. In our example, one could also investigate into the coldness of diamonds. We could think of a context where coldness is of relevance and then, the more precise degrees of the coldness of a diamond would indeed be a relevant part of the concept of a diamond. None of the possible propositions are relevant per se — they become relevant only in relation to a context where the practical bearings are played out.

Thirdly, the connection between effect and practical bearings seem to insinuate that this kind of research logic only works for the natural sciences and part of the social sciences because of the experimental setting Peirce suggests. Does that imply that science education research, which is mainly social science and humanities oriented could not use the pragmatic maxim? Although Peirce has not addressed this problem directly, one can nevertheless point to the possibility to replace the laboratory experiment with a much broader understanding of experiment. Then, also the science education research could use the pragmatic maxim as a relevance test. The strength of the pragmatic maxim lies in its ability to link pure causal effects of any intervention into reality with practical consequences that are of social, political, cultural or ethical nature. Therefore, the social sciences and humanities who are investigating and reflecting the social, political, cultural and ethical conditions, should be seen as a natural ally of almost any type of research that has practical consequences. This is also the case for science education research. Here purely didactical or even technical problems lead in a similar way to practical consequences that are of much broader scope.

A good example of how the pragmatic maxim works to help us to think of practical consequences is the case of preimplantation genetic diagnosis (PGD) for parents with a genetic risk that would affect their children in a most negative way. PGD is used to select embryos in IVF (in-vitro fertilization) and thereby to avoid selective abortion in a later stadium or a sick child. The technology raises ethical awareness, however, what precisely is the ethical relevance of this technology? What is ethically relevant here is not the technology as such, neither the fact that a child with specific characteristics will be born, but the practical indeterminacy of how this genetic selection of an embryo would affect our self-understanding as human beings and society. The indeterminacy in relation to the practical consequences is also what Jürgen Habermas (Habermas 2003) emphasizes in his analysis of PGD, namely that it might have effects on the way human beings think of themselves as not being fully autonomous (but a person of design) and thereby changing the basic framework of how we interact with each other. Will PGD change how we think of the status of human beings? Will we no longer think of the human being as an end-in-itself but as instrumental to other ends? The practical bearings of this technology are unknown and this fact poses severe questions to us. Ethical relevance cannot be found in the pure conceptualization of PDG as technology, but it is the possible practical bearings that shed a very distinct light on what the ethical problem actually is. To the same intimate relation between the conceptualization of a research subject and its practical context has Habermas (1994) pointed when he says that the truth that results from inquiry "is not derivable merely from logical rules of the process of inquiry, but rather only from the objective life context in which the process of inquiry fulfills specifiable functions: the settlement of opinions, the elimination of uncertainties, and the acquisition of unproblematic beliefs – in short, the fixation of belief." (Habermas 1994, p. 119) Irrelevant research has no practical consequences in the life world context. This way of looking at research encompasses necessarily uncertainty about the outcomes, but this is precisely what makes research worthwhile and relevant and, as Peirce and Dewey have emphasized, without real uncertainty research starts to be scholastic or even dogmatic.

Fourthly, the individual researcher can fail and make errors. This deficit "can only be corrected by the work of the whole ongoing community of inquirers"

(Haack 2018, p. 214) Being in a team of researchers the individual biases and idio-syncrasies are not enabled to flourish but are questioned and revised. Many research fields today, and science education research is one of them, afford both special expertise and cross-disciplinary teamwork. There are numerous examples for the success of these collaborations. However, it is not only cross-disciplinary teamwork that is necessary, also collaborations between researchers and activist groups or other citizen organizations have shown to be most fruitful (Bohman 1999; Frodeman 2014). One does not need to go as far as Peirce who reformulates the theory of truth as a process of inquiry: true is what the scientific community could agree upon in the long run – truth as a regulative idea of research practice. For the question pursued in this chapter it is important to direct our attention to the fact that truth and scientific results are not 'private' endeavors but the outcome of a collective process of deliberation, of evaluation, of mutual critique and correction.

The idea of practical bearings that Peirce brings forward seems to be similar to the type of relevance that stems from meeting practical needs of society. This is partly correct. Peirce, and even more so, Dewey think of research and inquiry as having the duty and ability to solve pressing problems. However, this does not necessarily imply that research – and this includes science education research – should naively follow given political or economic goals. The task of defining the problem cannot be delegated to the public, rather the researcher has the task to take up questions that have their source in reality and that lead to research that will have practical consequences. This understanding of relevance goes well with methodological pluralism. The pragmatists are not reductive when it comes to methods and research design. Peirce explicitly says "Don't block the way of inquiry" (Peirce 1998, p. 48) and hereby he meant to keep research free from dogmatism and allow a plurality of approaches flourish. The term 'practical bearings' Peirce introduces, reaches much broader than 'solving pressing problems' and indicates a general awareness for the social responsibility of research and not a purely economic or political interest: "It is to be remarked that the theory here given rests on the supposition that the object of investigation is the ascertainment of truth. When an investigation is made for the purpose of attaining personal distinction, the economics of the problem are entirely different. But that seems to be well enough understood by those engaged in that sort of investigation." (Peirce 1958, p. 157).

The beauty of Peirce's pragmatic maxim lies in the elegant solution for the problem of criteria for the relevance of research. To exceed the pure internal academic standards for knowledge accumulation we do not need to take our residue external to research. The practical bearings that shed a light on the object of our research are issues that stem from the societal or scientific context but are intrinsically connected to the research process by the researcher and the choice of hypotheses. The societal and life-world context is a condition for making research possible rather than its hindrance.

2.4   Relevance and Social Responsibility

We have seen that ethical concerns and epistemic considerations are not so neatly separated as classical philosophy of science wanted it to be (Tuana 2013). The boundaries are not so clear-cut anymore and the pragmatic understanding of research and its situatedness in society underpins this point. Values internal to research are necessary and inevitable for doing research. Dewey's approach to science and its responsibilities in relation to society tries to circumvent the opposition between objectivity and value-ladenness by viewing research as an intermediary process where practical contexts are the start and the endpoint of research. Problem solving is one, yet rather reduced, function of research. For Dewey, research and science enrich the world with new types of world views and thus are a part of a constant transformation of the way we approach and interpret the world. Ethics then becomes an integral part of research and is not imposed from outside as the work of ethics committees sometimes is perceived. Dewey's approach fits very well to the newer developments of integrating ethics in research. Frodeman's approach to sustainable knowledge and applied philosophy is a point in case. He suggests that philosophy, the social sciences and the humanities should engage with the academic and non-academic public by "integrating philosophical case-work into the daily tasks of the public and private sectors" (Frodeman 2014, p. 102). The production of knowledge in contexts of use challenges the traditional understanding of philosophy, the humanities, and the social sciences. Another case in point are initiatives and research approaches that focus on responsible research and innovation, a view on research that "implies that societal actors, such as researchers, citizens, policy makers, companies and civil society organisations work together in the whole research and innovation process in order to better align both the process and its outcomes with the values, needs and expectations of society" (SiS net n.d.). To a certain extent this recommended integration of citizens is already realized in science education research when pupils, school teachers and parents become more than the research object but an integrated part of conceptualizing the research question and a responsible part of the research process.

The forementioned approaches share with Dewey's understanding of research at least two main characteristics. Firstly, research is collaboration, often with interdisciplinary teams but also with the broader public. In his writings on education, Dewey highlights for example the role of the teacher as an investigator (Dewey 2008b, p. 23), thus pointing to a possible and fruitful collaboration between researchers and practitioners of education. This is also a new collaborative research practice that is used in science education research (Cai et al. 2018). Secondly, research is embedded in society and should neither be seen beyond nor opposed to society. Both characteristics together result in a very different way of how we perceive of research today compared with the ideal of research when the positivists developed their ideas of value neutralness. The model of the individual engineer is outdated. In this model the individual engineer has determinate knowledge about causes and consequences and he is in no need to justify or discuss his research with the public. This has changed tremendously and the public's right to say has been institutionalized in many countries and supranational institutions like the EU. A case in point are technology assessment councils and initiatives that include citizens' voice in their advice to politics, for instance the German Office of Technology Assessment at the German Bundestag.[6]

Are these developments that could roughly be characterized as the integrative approach to research ethics merely the reflexes of research institutions like universities or funding institution like the EU to support industrial research interests with the goal to keep the public quiet and affirmative to processes of technological innovation? From Dewey's point of view the answer would be a mixed one. Every good concept and model could be misused, he would argue. Pragmatism or ideas about social responsibility are no exception. However, he would argue that awareness of relevance of research and a better understanding of the nature of the research process as such, could help to avoid that research is dominated by political, commercial or specific social groups' interests.

In the remaining part of this chapter we shall see how Dewey conceptualizes the relevance of research and whether he successfully circumvents the danger that research and science is instrumentalized for short sighted goals and objectives that lack democratic legitimacy. Indeed, Dewey warned explicitly education research, a discipline that he thought was only in its infancy, to give in to "a pressure for immediate results, for demonstration of a quick, short-time span of usefulness in school." (Dewey 2008b, p. 8) He highlights that the context of education is highly complex and one should never implement a selective research result directly and without further reflection as a rule or new norm in the educational system. Dewey's article on the science of education is interesting for at least two reasons. Firstly, Dewey defends his understanding of science and inquiry against an understanding of the science of education that relies on allegedly natural intuitions and not on empirically corroborated knowledge. These are the same arguments he brought forward to foster a reconstruction in philosophy and a transformation of the broader context of the humanities. It is highly problematic for Dewey if research is incapable of putting its concepts and theories at work and thus testing them for possible effects that, in the words of Peirce, have practical bearings. Without such a relation to practical consequences science education research would confine themselves to narratives and suggestions that might not have much grounding in reality.

Secondly, Dewey highlights the value of scientific inquiry by making clear that the direct "transformation of scientific findings into rules of action" (Dewey 2008b, p. 9) is a misunderstanding of what science provides. Dewey addresses very directly the problem that stems from a notion of relevance that is reduced to what is relevant for the direct needs of society. The pressure these needs exert might be tremendous, however, the researcher has to bracket the research process in order to intelligently, and not instinctively, answer the demand. As a general rule, Dewey therefore states that the "preoccupation with attaining some direct end or practical utility, always limits scientific inquiry" (Dewey 2008b, p. 8). This is quite in accordance with

---

[6] https://www.tab-beim-bundestag.de/en/index.html

others who as Martin Carrier (2011) are concerned with the epistemic quality of research if direct utility rules. However, Dewey does not think practical relevance as such distorts the epistemic quality of research. He insists upon that research is intrinsically practical because it starts with a research problem that has its source in a real research situation. Only those research questions are relevant that would lead to practical consequences, contribute to problem reflection and if possible to problem solution. Starting and endpoint of research are real life contexts. It is not direct utility that rules research in Dewey's approach, and therefore epistemic quality is not threatened or compromised. He would, however, argue that epistemic quality is not an end itself but has the function to produce good research, that is research with practical consequences that contribute to a better society and better life conditions.

How then, does society and practical needs play a role for Dewey's understanding of the relevance of research? On a more general level Dewey argues that research and science are part of the greater experiment he calls democracy; both science and democracy build on cooperative and exploratory actions. This is also the way Dewey thinks of education, which has to be democractically organized and should emphasize cooperation instead of competition. Education, research process and democratic practices are characterized by intelligent cooperation to find solutions for the problems that are not pre-defined, but are demarcated and formulated against a public milieu (Dewey 1927, p. 157ff.). In this public setting scientific theories are challenged, rather than merely de-valued or subordinated under the premises of practices. Dewey's pragmatism aims to improve these practices through the democratic experimental community, cooperative education and the use of scientific methods. What Dewey aims at have others described as the role of science for the common good. Merton's CUDOS norms formulate clearly that 'communalism' is necessary for science and should be clearly distinguished from politicization and commercialization (Merton 1973). The common good is a much broader concern than economic or political interests which are always group related or even individualized. Science should be aimed at the common good, something that is in the public interest and not merely of economic interest (Radder 2017).

On a more specific level, research and relevance are for Dewey intimately connected because of the intertwinement of what Tuana (2013) called the intrinsic and extrinsic dimensions of ethics. In Weber's terminology we could also say that the guiding values (Wertideen) constitute a solid ground for research that is relevant to a society (Weber 1988). While Weber emphasizes the necessity of reflection upon these guiding values, for Dewey the role of hypotheses and their function for thinking of effects that have consequences in reality are more relevant. In this context it is worthwhile to have a look at Dewey's direct recommendations for the social sciences, which nicely sum up what from his point of view counts as relevant research: "In fine, problems with which inquiry into social subject-matter is concerned must, if they satisfy the conditions of scientific method, (1) grow out of actual social tensions, needs, 'troubles'; (2) have their subject-matter determined by the conditions that are material means of bringing about a unified situation, and (3) be related to some hypothesis, which is a plan and policy for existential resolution of the conflicting social situation." (Dewey 1986, p. 493). Applied to science education research we can say that the first criterion named by Dewey refers directly to society, to those practical contexts that bring about a situation in which the researcher is asked to react and start a process of inquiry, which will then hopefully result in an "existential resolution of the conflicting social situation". The first criterion has to be coupled with the two other criteria that address the research process as such. While the first criterion urges the science education researcher directly to think of how her research is related to needs of society, the second criterion refers to the research process proper: "An idea of an end to be reached, an end-in-view, is logically indispensable in discrimination of existential material as the evidential and testing facts of the case. Without it, (…) one 'fact' would be just as good as another- that is, good for nothing in control of inquiry and in formation and settlement of a problem." (Dewey 1986, p. 491). This quote can be interpreted as a critique of research that has lost its guiding value and is randomly producing facts and data without any real consequence. In Dewey's view relevance of research is not imposed from outside, but stems from the subject matter and the way in which the researcher handles this with ends-in-view and hypothesis at work. The relevance Dewey recommends is neither governed by political or economic interests nor by a misunderstood idea of research freedom that postulates that whatever topic the researcher chooses is relevant.

2.5 Conclusionary Remark

We have seen that the pragmatic concept of relevance speaks into a very contested and mixed field. However, it should become clear that relevance is indeed a matter of ethics conceived in its broader sense of contributing to the common good of a society. The pragmatists Peirce and Dewey are convinced that relevance of research cannot be found by conforming to the inner-academic idea of basic research that is free of all concerns that might have practical consequences for reality. However, the pragmatists would defend the ideal of freedom of research from direct interests and demands because this would inhibit the creativity that stems from a research process that starts with an indeterminate situation with the goal to achieve knowledge and control over possible outcomes. Peirce's formula directs our attention away from the research process as revealing nature and towards a practice-oriented concept of knowledge. Only the difference that makes a real difference, one could say, makes research worthwhile and relevant. To the question of relevance to whom, the pragmatists argue that research should be a way of enhancing our lives, making better choices for future generations and in general having a meliorative intention. This has often been called naïve however, compared to the pure relevance that might lie in exploring the nature of things for no reason, or for the reason of curiosity, the relevance of contributing to better life conditions seems to be more attractive and plausible. In science education the meliorative intention would address e.g. social equality and access to teaching, gender equality as well plurality of teaching methods and tools. Science education research could also work towards a better collaboration with teachers as well as reflect the role science education plays for a broader political and economic agenda. Pragmatists like Dewey and Peirce would not refrain from the overall value that science education as such give pupils and students unique tools to understand the reality they live in and to work for the controlled transformation of this reality. Science education research should then bridge the gap between a pure skill and knowledge oriented education on the one hand and ethical as well as social reflections upon the practical consequences of the sciences on the other.

In a way the pragmatic understanding of relevance is also old-fashioned as it holds on to the idea of a common good that is situated beyond partial interests, be it political or economic interests. For pragmatists the only way to withstand political abuse and the current economic demands for immediate application is to acknowledge this responsibility of science education research and to take up a responsible attitude towards a pluralistic and critical debate inside and outside academia. It is a truism that a mixed field of interests influences science education research and the way in which the subject matter is constituted. From the pragmatist's practical understanding of research it becomes clear that funding preferences, political or economic interests are external and highly problematic criteria; but also the direct interests and normative concerns of researchers and other stakeholders are not necessary justifiable criteria of relevance either. It is precisely this combination of knowledge that is in principle fallible and selective and that at the same time is deeply situated in a society and its practices that places considerable pressure on researchers today. Faced with possible errors and mistakes, also the science education researcher following Peirce and Dewey must accept the challenge of navigating in a contested field and still have to make an attempt to engage in research that is of significance.

References

Bohman, J. (1999). Democracy as inquiry, inquiry as democracy: Pragmatism, social science and the cognitive division of labor. *American Journal of Political Science, 43*(2), 590–607.

Bryman, A. (2008). *Social science methods* (3rd ed.). Oxford: Oxford University Press.

Cai, J., Morris, A., Hohensee, C., Hwang, S., Robison, V., & Hiebert, J. (2018). Reconceptualizing the roles of researchers and teachers to bring research closer to teaching. *Journal for Research in Mathematics Education, 49*(5), 514–520.

Carrier, M. (2011). Knowledge, politics, and commerce: Science under the pressure of practice. In M. Carrier & A. Nordmann (Eds.), *Science in the context of application* (pp. 11–30). Dordrecht: Springer.

Carrier, M., & Nordmann, A. (Eds.). (2011). *Science in the context of application*. Dordrecht: Springer.

Christians, C. G. (2005). Ethics and politics in qualitative research. In K. Denzin & Y. S. Lincoln (Eds.), *Handbook of qualitative research* (3rd ed., pp. 139–164). Thousand Oaks: Sage.

Denscombe, M. (2002). *Ground rules for good research*. Buckingham: Open University Press.

Dewey, J. (1927). *The public and its problems*. Athens: Swallow Press.

Dewey, J. (1986). *Logic: The theory of inquiry, later works 12, 1938*. Carbondale/Edwardsville: Southern Illinois University Press.

Dewey, J. (1988). *The development of American pragmatism, the later works 2, 1925–1953* (pp. 3–21). Carbondale/Edwardsville: Southern Illinois University Press.

Dewey, J. (2008a). *The middle works of John Dewey, volume 12, 1899–1924.* Carbondale/ Edwardsville: Southern Illinois University Press.

Dewey, J. (2008b). *The sources of a science of education, the later works 5, 1929–1930* (pp. 1–40). Carbondale/Edwardsville: Southern Illinois University Press.

Farthing, S. (2016). *Research design in urban planning.* Los Angeles: Sage.

Flyvbjerg, B. (2001). *Making social science matter.* Cambridge: Cambridge University Press.

Frodeman, R. (2014). *Sustainable knowledge. A theory of interdisciplinarity.* Basingstoke: Palgrave Macmillan.

Frodeman, R., Mitcham, C., & Sacks, A. B. (2001). Questioning interdisciplinarity. *Science, Technology and Society Newsletter, 126–127,* 1–5.

Haack, S. (2018). Expediting inquiry: Peirce's social economy of research. *Transactions of the Charles S. Peirce Society, 54*(2), 208–230.

Habermas, J. (1994). *Knowledge and human interest.* Cambridge: Polity Press.

Habermas, J. (2003). *The future of human nature.* Cambridge: Polity Press.

Hammersley, M. (1995). *The politics of social research.* London: Sage.

Hammersley, M. (2009). Against the ethicist: On the evils of ethical regulation. *International Journal of Research Methodology, 12*(3), 211–225.

Hollis, M. (1994). *The philosophy of social science.* Cambridge: Cambridge University Press.

Kourany, J. A. (2011). Integrating the ethical into scientific rationality. In Carrier, M.& Nordmann, A. (Eds.), *Science in the context of application* (pp. 371–386). Dordrecht: Springer,

Merton, R. (1973). The normative structure of science. In *The sociology of science. Theoretical and empirical investigations* (pp. 267–278). Chicago: University of Chicago Press.

Misak, C. (2013). *The American Pragmatists.* Oxford: Oxford University Press.

Mitcham, C. (2003). Co-responsibility for research integrity. *Science and Engineering Ethics, 9,* 273–290.

Nowotny, H., Scott, P., & Gibbons, M. (2001). *Re-thinking science. Knowledge and the public in an age of uncertainty.* Cambridge: Polity Press.

Payne, G., & Payne, J. (2004). *Key concepts in social research.* London: Sage.

Peirce, C. S. (1935). *Collected papers V, pragmatism and pragmaticism and scientific metaphysics.* Cambridge: Harvard University Press.

Peirce, C. S. (1958). *Collected papers VII and VIII, science and philosophy.* Cambridge: Harvard University Press.

Peirce, C. S. (1998). First rule of logic. In *Peirce Edition project, the essential Peirce: selected philosophical writings* (Vol. 2, pp. 42–56). Bloomington: Indiana University Press.

Pietarinem, A.-V. (2005). Compositionality, relevance, and Peirce's logic of existential graphs. *Axiomathes, 15,* 513–540.

Punch, M. (1994). Politics and ethics in qualitative research. In K. Denzin & Y. S. Lincoln (Eds.), *Handbook of qualitative research* (1st ed., pp. 83–97). Thousand Oaks: Sage.

Radder, H. (2017). Which scientific knowledge is a common good? *Social Epistemology, 31*(5), 431–450.

Rescher, N. (1976). Peirce and the economy of the research. *Philosophy of Science, 43*(1), 71–98.

Resnik, D. B. (2007). *The price of truth. How money affects the norms of science.* New York: Oxford University Press.

SIS net. (n.d.). Responsible research & innovation. In *Network of national contact points for science with and for society in horizon 2020.* http://www.sisnetwork.eu/rri/

Tuana, N. (2013). Embedding philosophers in the practice of science: Bringing humanities to the sciences. *Synthese, 190,* 1955–1973.

Weber, M. (1988). Die Objektivität sozialwissenschaftlicher und sozialpolitischer Erkenntnis. In *Gesammelte Aufsätze zur Wissenschaftslehre* (pp. 146–214). Tübingen: UTB.

Williams, M. (2016). *Key concepts in the philosophy of social research.* London: Sage.

Antje Gimmler (1961), Dr. phil. is Professor of Applied Philosophy at the Department of Culture and Learning at Aalborg University and director of the Centre for Applied Philosophy (CAF). Her research fields are social philosophy, philosophy of science, philosophy of technology and democracy. Her theoretical background stems from pragmatism and neo-pragmatism as well as Critical Theory. She has been member of international expert groups evaluating interdisciplinary research, a topic that also is a major research interest of Antje Gimmler. Her current research interest is to look into collaborative forms of research such as citizen science methods. Latest publications: The normativity of culture, in: *Bringing Culture Back In – Human Security and Social Trust*. ed. M. Bøss. 1. ed. Aarhus: Aarhus University Press, 2016. pp. 121–134; Recognition: Conceptualization and Context, in: H. Leiulfsrud/P. Sohlberg (eds.), *Concepts in Action*, Leiden: Brill Publishers, 2018, pp. 302–320; Practices we know by – Knowledge as transformative, in: A. Buch/Th. Schatzki (eds.), *Questions of Practice in Philosophy and Social Theory*, London: Routledge, 2018, pp. 176–195.

Chapter 3
Reflections on Research Ethics in Historically Oriented Science Education Research in Canada

John Allison and Michaela Vogt

3.1 Introduction

In recent times, discussions of ethics have become much more prevalent in academia and popular culture, so much so that representations of the term have become common place. These reviews are an assessment of the integrity of research; in thinking about research ethics we are more often than not considering research that addresses live human beings and/or other living creatures. This research is sometimes scientific in nature, quite often quantitative, and frequently qualitative in nature. It can also ask direct questions of research participants. In the past, health care research frequently provided ethical review protocols for all different types of research as van den Hoonard notes (van den Hoonaard 2011). These biomedical protocols continue today to ensure that all risks to subjects are weighed and considered before the research is conducted. That these health-related protocols and approaches are an important part of research ethics is clear. In recent times however, new directions have emerged in research ethics in the field of science education. As the then new editor of *Research in Science Education*, Stephen Ritchie, presciently noted in 2008.

> Traditional stances on research ethics in education followed medical models that emphasized informed consent and privacy (e.g., Howe and Moses 1999). As researchers strive to enhance the impact of their work on students and teachers of science, I would expect a movement away from conservative methodologies. Accordingly, the relationships between researcher and other research participants will become more challenging. (Ritchie, 2008, pp. 1–2.)

J. Allison (✉)
Schulich School of Education, Nipissing University, North Bay, ON, Canada
e-mail: johna@nipissingu.ca

M. Vogt
Fakultät für Erziehungswissenschaft der Universität Bielefeld, Bielefeld, Germany

Along with the changing models for science education ethics review, we want to introduce historical research since we believe that interesting aspects to do with research ethics can be gained by examining this field. Thus, we focus particularly on research ethics in relation to historical methodology. This differs from the biomedical approach and directs us to look to the past and find sources of past events; things that may have happened in the last ten years, the last century, the last millennium or beyond. In the fields of science and science education, this type of research provides a link to past developments in the classroom and in the field. Scholars such as Zeller and Simon have shown the way in their work on the history of science (Hoffman 2013; Simon 2013; Zeller 2000, 2017).

Concomitantly, when historians look at primary source documents, they must do so with a somewhat skeptical and critical eye. As science education researchers we must ask ourselves then; are biomedical models and approaches to research ethics also a fit for this field as we undertake historically oriented science education research? This chapter proposes to answer this question in the following way: it will argue that the biomedical model of research ethics review is not appropriate for this type of research. In support of this view, the chapter examines research ethics review policies, definitions and protocols in science education in the context of Canada. The reason for examining the historical context of Canadian science education research is exemplary. Each country presents its own research histories and to deconstruct them allows identifying very unique trajectories. However, this development is also set in the greater context of historically oriented research in science education since research is an international endeavour. The chapter will subsequently discuss the specificities of historically oriented research in this field and delineate the different kinds of research data sources, whether these are primary source documents, or data from science education projects. The chapter will conclude with a reprise of the discussion about biomedical ethical review in science education research that is historical in orientation.

3.2 The Current State of Research Ethics Review in Canada and the Literature

3.2.1 Current State

When trying to gain further understanding of the biomedical perspective key questions need to be addressed in terms of establishing the current state of research ethics review in Canada and understanding the research ethics rules in science and humanities disciplines. These are the following: what is the overall existing framework for research ethics review in Canada? Also, what is the perspective of the literature on this topic? What role has the biomedical model played in this?

To understand the influence of the biomedical model, one must start with the structure of research ethics governance in Canada. The policy oversight at the highest level is comprised of the Three Councils. The more commonly named "Tri-Council" governs ethical research practices and the responsible conduct of research in Canada (Panel on Responsible Conduct of Research 2016). This is a federal government organization that consists of the Social Sciences and Humanities Research Council (SSHRC), the Natural Sciences and Engineering Research Council (NSERC) and the Canadian Institutes of Health Research (CIHR) (Canadian Council of Academies & C.C.A. Expert Panel on Research Integrity 2010). In recent years, there has been a worldwide trend to tighten research integrity and this has been the case both in Canada and in other advanced industrialized states such as Germany (Canadian Council of Academies & C.C.A. Expert Panel on Research Integrity 2010; Mayer and Steneck 2012; Steneck et al. 2015; Zimmerman 2015). Under the auspices of SSHRC between 1994 and 1998, the "Tri-Council Policy Statement: Ethical Conduct for Research Involving Humans" (abbreviated to TCPS) was created. It is a guide for university-based Research Ethics Boards (REBs) to use in terms of deciding whether research is ethical (Gontcharov 2011; Heslegrave 2012; Janovicek 2015). The original TCPS came about with the biomedical model of review very much in mind. In the history of research ethics in the United States this model originated with a focus on medical patient privacy. As Schrag defines it, there are a variety of foci of institutional review boards that reinforce the biomedical model. He speaks of four key components in defining biomedical model research: "1. Researchers know more about their subjects' condition than do the subjects themselves. 2. Researchers begin their work by spelling out detailed protocols explaining what hypotheses they will test and what procedures they will employ to test those hypotheses. 3. Researchers perform experiments designed to alter subjects' physical state or behavior, rather than simply gathering information through conversation, correspondence, and observation. 4. Researchers have an ethical duty not to harm their subjects. (Schrag 2010)." Ells and Gutfreund also provide a succinct definition of the biomedical model; "The 'biomedical model' for research is hypothesis driven; that is, researchers generally begin with a formal hypothesis, make use of experimental designs and quantitative data, and engage in deductive reasoning with the aim of confirming the hypothesis (Ells and Gutfreund 2006, 370)."

This focus on a biomedical approach and hypothesis testing resulted in a revolt in the early 2000s on the part of social science researchers, educationalists, humanities researchers and historians in Canada (Janovicek 2015). One of the organizations that led this revolt was the Canadian Association of University Teachers (CAUT) and they took exception to some of the power of collectives over research projects as Grant notes (Grant 2016). One area that these scholars took exception to was the area of privacy. As Canadian historians Franca Iacovetta and Wendy Mitchinson state, "our legal obligations as researchers to protect the privacy of individuals in the past can lead us to write the marginal into history by writing their names and faces out of it (Iacovetta and Mitchinson 1998; Janovicek 2015)." Consequently, in this regard, the difficulty of writing history, and dealing with issues of privacy and confidentiality remains. In contemporary times, groups of Canadian historians have continued to be attuned to ethical issues and have made statements with regard to how ethical practice plays a role in their research, notably, for example, the Canadian Historical Association (CHA) (Canadian Historical Association 2018). Additionally, as van den Hoonaard (2011) underlined in his book length examination of ethics review, the gap between the biomedical model and social science research is often extreme. He underlined this in his exposé of three case studies that looked at ethnographic research of the deaf-blind, street people and nursing homes; research that was not anonymous and relied on knowing the subject of the research (van den Hoonaard 2011, p. 63). Knowing the subject of the research was critical to the successful completion of the research in this circumstance. TCPS 2 in its current iteration came out in 2010 (updated in 2014 with a companion document) in an attempt to address the issue of the biomedical approach (NSERC and SSHRC 2014). While somewhat successful in speaking to the question, it still has its critics (van den Hoonaard and Tolich 2014).

Earlier, science education researchers also looked at the initial iteration of the TCPS and found it lacking in several key areas. These researchers were very critical of two main issues. Firstly, they disagreed with the original one-size-fits all, overly risk-aversion to legal exposure approach of the TCPS and the early institutional REBs. Additionally, they did not like the deeply probing questions that called into question disciplinary research designs. This research group concluded that science education researchers needed to be more proactive in the development of research ethics policy and review; otherwise, the *big-stick* approach to research ethics was unlikely to change in the near future (this was the view in 2008, TCPS was revised, as noted, in 2010 and 2014) (Anthony et al. 2009).

Notably as well, there is increasing recognition by science educators and academic communities that more must be done in terms of educating for ethical process and the correct application of policies and programs such as TCPS 2 as Stockley, et al. note (2016). As ethical review will be with every scholar in the field of education moving forward, ethical education must be addressed as part of one's doctoral education and must be explicitly mentioned in one's dissertation. Scott's work is one early example that illustrates the greater awareness and use of research ethics protocols by doctoral students in the early twenty-first century (Scott 2007).

3.2.2 The Contributions of Different Scholarly Literatures to the Dialogue on Educational Research Ethics Policies and Protocols

In addition to the above-mentioned group of historians, it is important to examine whether scholars from several disciplines and research fields have looked critically at research ethics review and questioned why use the biomedical model. This section addresses highlights of some of their perspectives. As these research fields do also contain historical perspectives it is important to include their viewpoints as well.

The biomedical approach and research ethics review has not been well discussed in the field of the history of education. The evidence of reflection is slim. In her recent article on research "North of 60", McGregor (2015) speaks to the question. Particularly, in her case, the challenge of research in Canadian Inuit communities was that it required ethical approval from several levels of government, Indigenous communities, and institutional ethics boards and the requirements for these ethics boards were often at odds (McGregor 2015; c.f. Nickels et al. 2006). Raptis (2010) in her encyclopedia entry on documentation notes the problematic nature of restricted access documents. Archival personal records that are restricted must be anonymized before they can be used.

In the field of Indigenous studies, questions of research ethics review and appropriate types of protocols present a more lively discussion. Research in this field is often at odds with a scientific/biomedical model. Scholars in this field look at questions about First Nations, and the Indigenous peoples of Canada, including their history. Following the very horrific and tragic historical events of Residential Schools, the more recent establishment of the Truth and Reconciliation Commission (TRC) and the Aboriginal Healing Foundation have focused a strong light on research ethics and Indigenous people in Canada (Battiste 2008; DeGagné 2012). While this is a topic too broad to cover in this analysis, it is important to signal its significance.

Scientists and science education researchers around the world and in Canada have also become increasingly concerned about ethical considerations and integrity in their teaching, research and data collection (Schoenherr and Williams-Jones 2016). There has been an awareness over a long period of time that the direction of science education research is important and that science education can have deep impacts on culture and what science teachers do in the classroom shapes societal views (Bazzul 2013; Frazer and Kornhauser 2014). In more recent times science education researchers have discussed ethics in a variety of contexts as Bazzul notes (2016, pp. 24–26). In the early 2000s, science educators were also concerned about the broader implications of their teaching and the ethical responsibilities of science graduates as well as their need to be schooled in the broader socio-political currents that impact the scientific endeavour (Hodson 2010). Nielsen echoes this in his discussion of "supplementing regular science teaching with socio-scientific issues; that is, social, ethical, and political issues that have conceptual ties to science (Nielsen 2013, 373)." In his view, how students articulate evidence in socio-scientific discussion is equally important to science knowledge (Nielsen 2013).

In contemporary times as well, the editors of the primary journal or as the editors describe it, the "mouthpiece" of science education research in Canada, *The Canadian Journal of Science, Mathematics and Technology Education* have retrospectively reflected on the direction of science education research in Canada. They note the various approaches to research; quantitative, mixed-methods (of which the contributions are few), and qualitative research including action research, ethnography, theory building, interpretation, and narrative based research (Pegg et al. 2015). Less was said however, about the ethical implications of science education research in the editorial reflection. Ethics was examined as part of classroom practices as opposed to research (Pegg et al. 2015). Other new works on science education in Canada speak to teaching ethics in the classroom, but not the implications ethics review for research (Tippett and Milford 2019). With the paucity of reflection in this area, it signals the need for more and new questions. This is a good opportunity then to look at historically oriented science education research, some of the challenges imposed on this field by the biomedical model of research ethics review, and how this fits into the broader picture of research ethics models.

With exception of Indigenous studies, research ethics protocols and research ethics review boards have seen only limited commentary by science educators and other scholars in Canada. This lack of commentary is in part because of the focus on "doing" research in their fields, as opposed to looking at it more obliquely through a research process focused lens. van den Hoonaard and Hamilton's (2016) work are the almost singular exception to the lack of critical analysis of research ethics review processes in Canada. Only a few scholars have explored this area, and this may be the case because critical reflection on practice is not often done. Additionally, for many, research ethics applications are a relatively new experience.

3.3   Specifics Regarding Research Ethical Standards

3.3.1   Science Education and Historically Oriented Research Methods and Archival Sites

Having examined the selected literature and the contexts of ethically based research in science education research and in other fields in Canada, we turn now to exploring the practice of data collection by utilising archives and the archive itself. It is clear from the earlier discussion that the biomedical model is difficult to apply to science education research. This exploration then looks first at science education research methods. Then, it turns to historically oriented methods, the use of archives, and the use of primary source documents.

3.3.1.1   Qualitative Research and Science Education

Distinct from biomedical research methods, science education researchers use qualitative and quantitative methods, such as surveys and interviews as some of the methodologies in the conduct of their research. Vogrinc and his colleagues describe in depth the wealth of methodological approaches that are currently used in science education research (Vogrinc et al. 2019). They write that the issue of data collection is also important, since:

> ... data collection [one] is not limited to one source or one technique only. Apart from the data acquired by interviews and observation, usually also different documentary sources are used, such as personal documents (a birth certificate, an employment record, a passport,

letters, photos ...), different records produced in the process of data collecting, transcriptions of tape recordings, video material, etc. It is important emphasize that only the triangulation – the pluralism – of data collection techniques and their mutual combination can provide for linking the findings of individual phenomena or aspects into a meaningful integrity (Vogrinc et al. 2010, p. 2).

The data is then coded, and this is central to the analysis of documents as Bryman notes (Bryman 2004). While Vogrinc and his colleagues write about personal data, in the science classroom and science education research this can be expanded to "classroom observations and artifacts" as data as Carrier notes in her study of scientific literacy (Carrier 2013, p. 11.). Further, rich research possibilities exist for the analysis of students' scientific vocabulary, their fluency with scientific terms, and their growing awareness of socio-scientific issues in the elementary, and secondary classrooms (Leeman-Munk et al. 2014, p. 4).

It is the researcher's responsibility to protect the individual and to uphold high standards of practice in science education research. Vogrinc and his colleagues conclude that science education researchers have a variety of ethical issues to address in their research:

> The general issues that need to be thought of are: informed consent, confidentiality, avoiding harm, integrity and professionalism. In research, ethical issues must be considered at each step in the research process. Ethical principles dictate: (1) what measurement techniques may be used for certain individuals and certain behaviors, (2) how researchers select individuals to participate in studies, (3) which research approach may be used with certain populations, (4) how studies may be carried out with individuals, (5) how data are analyzed, and (6) how results are reported (Vogrinc et al. 2010, p. 6).

These principles are very important for research with human subjects, but the question remains how applicable are they for historically oriented research? Which ethical approaches are significant here?

Historical case studies into the history of science and science education topics are where the question of ethics and the suitability of the bioethical model of research comes to the fore. For science education researchers, looking at the history of science and the history of science education through case studies is essential as these studies provide other views into potential ethical problems in their research and the teaching of research ethics (D'Angelo 2012; Pimple 2007). For example, in Herrid's work, several case studies are presented. These include "A Rush to Judgement," about a psychology lab and research protocols, through to "Bad Blood: The Tuskegee Syphilis project," a historical examination of an unethical clinical study using volunteers from Tuskegee University between the 1940s and 1970s (Herreid et al. 2012). Similarly, in the Canadian context, the analysis of cases on research ethics is a critical first step for researchers doing historically oriented science education research.

The question of ethics and the suitability of the bioethical model of research is additionally present for science education researchers who are working on historical biographical studies of science educators, policy analyses of science education, and historical research on the teaching of science and science curriculum in the classroom. In terms of biographical case studies, Hankins in an early article writes about the evolution of history of science biographical case studies (Hankins 1979). One rich source of curriculum documents regarding the history of science education is the Ontario Historical Education Collection at the University of Toronto. It provides primary source science education curriculum documents and primary source assessment documents as well (University of Toronto Libraries 2019).

Research ethics in historically oriented research is additionally closely tied to the evolution of archives. But, from the outset much of the biomedical review model is unsuited for these archives and archival searches as this evolution is not in any way tied to generating experiment designs, or the proof of hypotheses. Rather, archives in recent years underwent an upheaval driven by archivists and researchers who examine archival policy. In Australia for example, there is a discussion of "archive mania" and Derrida's "archival fever" (Biber and Luker 2016). Additionally, outside academia, lawyers among others, have become much more attuned to the use of archival sources and in re-examining old cases long forgotten but now resurrected with the use of archival records (Biber and Luker 2016). Much debate regarding the role of ethics as it is linked to archival documents is thus very relevant in contemporary times. McNeil notes that the archivist's role has been reimagined not only to serve as a trustee of documents and to ensure the completeness and security of the initial documents following their accession to the archives; but also to create archival codes of ethics, to protect of the privacy of individuals, and to protect their lives with regard to public exposure (MacNeil 1991). The refocus in the direction of privacy in the archival space is of critical import to an understanding of research ethics in science education in Canada. This focus on privacy does not directly speak to the biomedical model of research ethics review, it is important to see this as an adjunct and an evolution in archival thinking.

Further, it is important to understand the ostensible "archival divide" when we look at the applicability of biomedical research ethics protocols to historically oriented research (Blouin and Rosenberg 2012). This divide between researchers and archivists complicates questions of research ethics and historically oriented science educational research. The history of archives, their position of authority, their historically intense secrecy, their role in questions of diplomatics, and in general their role in societies, is long, dating back to medieval Europe and before (Blouin and Rosenberg 2012).[1] Archives and archivists, as noted earlier, defined new missions as the twentieth century wore on. The old mantra of "keep everything" gave way to new guiding principles. Instead, key questions now include what to collect, what to keep, when can people see the documents, and who should see the documents (Blouin and Rosenberg 2012)? Additionally, who would or should share the documents, as Nelson notes in his piece (Nelson 2009). One thing that was increasingly clear was that the integrity of archival records, and subsequently having a code of ethics associated with this, was absolutely critical (Cappon 1982; Nelson 2009). Further, the question of "enduring value," became one that archives, and archivists

---

[1] N.B. Diplomatics is the study of the veracity and authenticity of works as opposed to diplomacy. Diplomacy is the day to day conduct of international relations.

asked regarding primary source documents. Which documents should be kept? Archivists and historians also need to probe the dark side of archival policy, historical research and writing and look at a further question; which documents should be destroyed (Blouin and Rosenberg 2012)?[2]

Additionally, this divide bears added scrutiny in terms of research ethics as archivists move in the direction of a more quantitatively centered, potentially more biomedically oriented approach, greater scrutiny in records management, and greater integrity of the records (de Chadarevian 2016). There is a recognition in the archival community that historians and those undertaking historically oriented research constitute, still to this day, the major users of their resources (Anderson 2004). After surveying archivists and historians, Anderson, in his study, demonstrates that historians' information gathering techniques are also becoming more sophisticated and diverse. Additionally, historians' and those undertaking historically oriented research expectations around the detail of finding aids are becoming more challenging. These scholars are demanding more at a distance information that they can access through the Internet and the World Wide Web (Anderson 2004). Simultaneously, as noted however, in terms of research ethics review, archivists are also becoming more restrictive in what collections are available to the users of their collections. The whole of the archive is no longer the territory of the historian (if it ever was), many things are now heavily restricted and frequently require access to information requests in order to be examined and read.

3.3.2 Data Sources, Types of Information and Ethics

Having addressed the challenges concerning archives it is important now to look at research data, in this case primary sources and how the biomedical review model is applicable (or not) to this data. Primary sources constitute the nuts and bolts of historically oriented science education research. These documents, notably written documents and the photographs, are very important in determining the quality of the writing and research that will emerge. In this section we look at some of these sources. These have a growing impact on the kind of data that are now looked at by researchers. Therefore, we now turn to examining this data.

---

[2] This is done generally under a view to prune collections and keep that which is archival eye most significant. The more extreme version of book burning is another issue throughout the twentieth century as well. Sometimes due to war and insurrection, the archives are displaced as Lowry amongst others notes in his collection of essays. And archives can drag in the unsuspecting historian.

3.3.2.1 Written Sources

Historical data, as opposed to primary data collected through qualitative or quantitative methods (e.g. observations, interviews, surveys), comes primarily in the form of documents and these are central to the majority of historical writing projects (Bombaro 2012). In many ways it is similar to some of the other documentation science education researchers would collect using qualitative methodologies. As Bombaro and other historians have noted, historical primary source documents are bits of evidence including diaries, journals, government documents, artifacts, and images that are created firsthand by the person witnessing the historical event taking place (Bombaro 2012). Several issues arise with primary source documents. These will be briefly examined, but it is clear that a biomedical approach to ethical review might not be the right fit for this data.

The interpretation of historical documents is one challenge. Biomedical review models are challenged in this environment. Firstly, there is a possibility of bias with the interpretation of documents. When authors and historians examine primary source documents as McCullagh notes, bias can appear in a variety different way in the writing of historical documents, including through historical inference, historical explanation, historical description and historical interpretation (McCullagh 2000). Biomedical review models generally are not receptive to these approaches with their emphasis on formal hypotheses, experimental design and quantitative data. McCullagh additionally notes that some theorists feel that bias is unavoidable in sources. Ethically speaking this poses challenges. He argues further that the views and meanings of historians direct every aspect of the explanation of events in the past (McCullagh 2000). Consequently, for some theorists, this means the end of history as a discipline. Additionally, central is the notion that historians choose primary sources that interest them and consequently there can be no better, nor no worse representations of the past (McCullagh 2000). McCullagh also states that primary source documents simply need to speak on their own. The more voices from the past on a topic; the more perspective one gets, and a better picture of past events is thus rendered (McCullagh 2000). In terms of applying the biomedical review model here, it does not fit to these issues.

Further to this challenge, there is the challenge of multiple voices in many sources. Historians' writing and how primary sources are viewed should instead be seen through the presentation of opposing views. This contrasts with the idea of having simply a consensual view of the past as presented through primary source documents (Burke 2001; McCullagh 2000). In Burke's view, having a multiplicity of sources potentially eliminates the question of bias (Burke 2001).

3.3.2.2 Photographs

Photographs and films as historical data sources have been used in science education classrooms and the application of the biomedical ethics review model to these sources is inappropriate (Kafai and Gilliland-Swetland 2001). Science education researchers undertaking historically oriented investigations using these materials also face the prospect of ethical review. Dussel's and also Daston and Galison's works underline that photographs and visual media have their own set of ethical questions and issues. Daston and Galison in particular look at objectivity in the making of scientific images (Daston and Galison 2010; Dussel 2013). The scientist/ photographs' ability to present data in the photographic medium is often beset with issues that derive additionally from the nature of the "objective" devices capturing the image. Also, historical photographs are challenged by questions of interpretation and this is what the discussion moves to next.

Archival preservation of photographs and films has also become also increasingly critical as these media age. As Blouin and Rosenberg note, there are many questions around photographs. Central to these questions are issues such as the nature of the subject. Was it captured *as such* naturally? Or was it staged? What was the nature of the equipment used? What was the photographer trying to convey (Blouin and Rosenberg 2012)? Photographic archivists have long argued that archives need to look more closely at photographs and ask many questions (Blouin and Rosenberg 2012).

Questions of ethics and privacy are among the factors that were not questioned prior to the 1960s when photographs were used as data taken by science education researchers undertaking historically oriented research. Tinkler refers to this in her work;

> In recent years there has been a dramatic shift in perceptions of the rights of individuals to privacy. Fifty years ago, when Townsend undertook his study of residential care for the elderly, he encountered no restrictions and his photos were taken with little regard for the self-respect or wishes of the elderly people he depicted. This was not unusual at the time; documentary forms of photography, including academic studies, have a history characterized by a lack of regard for the people that are photographed (Tinkler 2013, p. 196).

Ethical reviews also come to the fore in terms of legal restrictions on photographs. Darling (2014), in her article on this topic, addresses the importance of researchers and others preparing for what may come in terms of graphic photographs (i.e. scenes of intense violence, death, etc.). While science education researchers in all probability will not use such images, it is important to know about these types of pictures. Darling's experiences underline the difficulties of seeing photographs without adequate preparation and without a sense of the context (Darling 2014; cf. Maynard 2017). Photographs can also be altered. Photographs help construct reality. Photographs can also be ambiguous. The more layers a photograph possesses, the greater this ambiguity and concomitantly, the greater the challenge of the researcher to accurately analyze and present the photographs as a primary source documents (Evans et al. 2018).

Two issues are very important; the providence of photographs, and the documentation of photographs (Vervaart 2014). Vervaart writes on a series of issues that science education researchers need to take into account in all types of research, historically oriented or otherwise; what equipment was used in obtaining the photo, why was a particular image selected, were there alterations made to the image, what are the details of the image, were annotation tools used to explain the image, and lastly, what are the details of the analysis (Vervaart 2014). Additionally, as Chouliaraki and Blaagaard note, the relationship between the researcher, the photographer and the image are critical (2013). Moreover, from the standpoint of the Canadian Tri-Council, issues of privacy and the sensitivity of the content of the photograph are also important in evaluating an ethical review protocol for photographic documentation (SSHRC Ethics Special Working Committee 2008).

Many of these questions come to the fore in Bullock's work (2014). Bullock, in his self-study of teacher education, as a historian-physicist researcher, talks about the analysis of video data using different approaches, the lens of the viewer, the lens of the researcher and the lens of the science teacher educator. While ethics was not a central part of his analysis, he noted "that the use of video in my teacher education classroom has been worth the additional ethical complexity of completing such a study (Bullock 2014, p. 45)."

3.4   Conclusion

In this chapter, we have argued that the biomedical model of research ethics is not appropriate for historical research undertaken in the field of science education. Through defining the biomedical research model, it is clear that it is very hypothesis driven and applicable to the health sciences. It is equally clear that national research organizations (i.e. SSHRC) and institutional research boards have taken a long time to distinguish and differentiate between different types of research for which ethical review would be necessary. To their credit, in recent years they are making some progress that would satisfy some of the critiques of the biomedical approach. Yet, there is still more to be done. In understanding the difference between a hypothesis driven framework and historical research, it is helpful to examine some of the sites of data and the nature of historical primary source data. In terms of science education research, the origin, the nature, and the type of source data is very important to science education researchers undertaking historically oriented research. Given the nature of data, biomedical research ethics principles are very difficult to apply in this field. Data collection is significant and requires triangulation regardless of which research method the science education researcher follows. In general, however, data sources such as primary sources are not able to be easily assessed using a hypothesis testing, experimental design approach as advocated by many biomedically inspired research ethics review boards as Schrag (2010) describes them. Instead, historical researchers let the documents speak. Using the example of Canadian science education research, we found nested and historically traceable practices. Under the current regime of research ethics in Canada, there is a very strong emphasis placed on making sure that individual privacy and confidentiality is preserved. In many contexts this is of great advantage in the sense that as individuals, Canadians do not want their personal data exposed. At the same time, it also raises issues that in terms writing scholarly history, as Iacovetta and Mitchinson (1998) note; the marginalized are simply eliminated from these histories as they are never identified. The nature of sources; primary source documents, and access to restricted documents are critical in this regard, bearing in mind the rights of each individual to privacy. Each of these different types of sources comes with an array of challenges and areas where sensitivities are important in terms of research ethics.

Looking then at these arguments we can in the end conclude that especially the issue of how research ethics is assessed and what protocols are used is vitally important to science education researchers in Canada and worldwide. That relying on these protocols in general is reasonable and beneficial has at no point been doubted in this chapter. But in order to do so, these protocols need to fit to the specifics of science education research in all its facets – including historical methodology. Therefore, the biomedical model is important as one example and definitely one not to follow, but to learn from, as science education researchers strive to develop excellence in their own research ethics protocols. It is also important to see the potential in the history of science and the history of science education research in this regard. It is important to open up science education to further historical research. This chapter has argued for more work to broaden our horizons for instance through the unique insights stemming from historical research. The investigation and conduct of more historical studies in this area of research is very much necessary going forward. There is still much work to be done.

References

Anderson. (2004). Are you being served? Historians and the search for primary sources. *Archivaria, 58*, 81–129.

Anthony, R. J., Yore, L. D., Coll, R. K., Dillon, J., Chiu, M.-H., Fakudze, C., et al. (2009). Research ethics boards and the gold standard (s) in literacy and science education research. In *Quality research in literacy and science education* (pp. 511–557). Dordrecht: Springer.

Battiste, M. (2008). Research ethics for protecting indigenous knowledge and heritage: Institutional and researcher responsibilities. In N. K. Denzin, Y. S. Lincoln, & T. S. Smith (Eds.), *Handbook of critical and indigenous methodologies* (pp. 497–510). Los Angeles: Sage.

Bazzul, J. (2013). *How discourses of biology textbooks work to constitute subjectivity: From the ethical to the colonial.* Toronto: University of Toronto.

Bazzul, J. (2016). *Ethics and science education: How subjectivity matters.* New York: Springer.

Biber, K., & Luker, T. (2016). *Evidence and the archive: Ethics, aesthetics and emotion.* Philadelphia: Routledge.

Blouin, F. X., & Rosenberg, W. G. (2012). *Processing the past: Contesting authority in history and the archives.* Oxford: Oxford University Press.

Bombaro, C. (2012). *Finding history: Research methods and resources for students and scholars.* Lanham: Scarecrow Press.

Bryman, A. (2004). Qualitative research on leadership: A critical but appreciative review. *The Leadership Quarterly, 15*(6), 729–769.

Bullock, S. M. (2014). Self-study, improvisational theatre, and the reflective turn: Using video data to challenge my pedagogy of science teacher education. *Educational Research for Social Change, 3*(2), 37–50.

Burke, P. (2001). *New perspectives on historical writing state college.* Pennsylvania State University Press.

Canadian Council of Academies, & C.C.A. Expert Panel on Research Integrity. (2010). *Honesty, accountability and trust: Fostering research integrity in Canada*. Ottawa: Council of Canadian Academies.

Canadian Historical Association. (2018). *Statement on research ethics | Canadian Historical Association*. Retrieved from http://www.cha-shc.ca/english/about-the-cha/statement-on-research-ethics.html

Cappon, L. (1982). What, then, is there to theorize about? *The American Archivist, 45*(1), 19–25.

Carrier, S. J. (2013). Elementary preservice teachers' science vocabulary: Knowledge and application. *Journal of Science Teacher Education, 24*(2), 405–425.

D'Angelo, J. (2012). *Ethics in science: Ethical misconduct in scientific research*. Philadelphia: Taylor & Francis.

Darling, J. (2014). On viewing crime photographs: The sleep of reason. *Australian Feminist Law Journal, 40*(1), 113–116. https://doi.org/10.1080/13200968.2014.931904.

Daston, L., & Galison, P. (2010). *Objectivity*. New York: Zone Books.

de Chadarevian, S. (2016). The future historian: Reflections on the archives of contemporary sciences. *Studies in History and Philosophy of Science Part C: Studies in History and Philosophy of Biological and Biomedical Sciences, 55*, 54–60.

DeGagné, M. (2012). *Speaking my truth: Reflections on reconciliation & residential school*. Ottawa: Aboriginal Healing Foundation.

Dussel, I. (2013). The visual turn in the history of education: Four comments for a historiographical discussion. In T. Popkewitz (Ed.), *Rethinking the history of education: Transnational perspectives on its questions, methods and knowledge* (pp. 29–49). New York: Palgrave Macmillan.

Ells, C., & Gutfreund, S. (2006). Myths about qualitative research and the tri-council policy statement. *The Canadian Journal of Sociology/Cahiers canadiens de sociologie, 31*(3), 361–373.

Evans, J., Betts, P., & Hoffmann, S. L. (2018). *The ethics of seeing: Photography and twentieth-century German history*. New York: Berghahn Books.

Frazer, M. J., & Kornhauser, A. (2014). *Ethics and social responsibility in science education: Science and technology education and future human needs*. London: Elsevier Science.

Gontcharov, I. (2011). *REB mission creep and positivism: TCPS and the problems of regulatory innovation in non-biomedical academic research. (LLM)*. Toronto: University of Toronto.

Grant, D. (2016). *When research is a dirty word: Sovereignty and bicultural politics in Canada, Australia and New Zealand ethics policies*. Otago: University of Otago.

Hankins, T. L. (1979). In defence of biography: The use of biography in the history of science. *History of Science, 17*(1), 1–16.

Herreid, C. F., Schiller, N. A., & Herreid, K. F. (2012). *Science stories: Using case studies to teach critical thinking*. Arlington: National Science Teachers Association.

Heslegrave, R. (2012). Research integrity in the Canadian context. In T. Mayer & N. Steneck (Eds.), *Promoting research integrity in a global environment* (p. 37). London: World Scientific.

Hodson, D. (2010). Science education as a call to action. *Canadian Journal of Science, Mathematics and Technology Education, 10*(3), 197–206. https://doi.org/10.1080/14926156.2010.504478.

Hoffman, M. (2013). Shunning the bird's eye view: General science in the schools of Ontario and Quebec. *Science & Education, 22*(4), 827–846. https://doi.org/10.1007/s11191-012-9517-x.

Iacovetta, F., & Mitchinson, W. (1998). *On the case: Explorations in social history*. Toronto: University of Toronto Press.

Janovicek, N. (2015). Oral history and ethical practice after TCPS2. In K. Llewellyn, A. Freund, & N. Reilly (Eds.), *The Canadian oral history reader* (pp. 320–330). Montreal/Kingston: McGill-Queen's University Press.

Kafai, Y. B., & Gilliland-Swetland, A. J. (2001). The use of historical materials in elementary science classrooms. *Science Education, 85*(4), 349–367.

Leeman-Munk, S. P., Wiebe, E. N., & Lester, J. C. (2014). *Assessing elementary students' science competency with text analytics*. In Paper presented at the proceedings of the fourth international conference on learning analytics and knowledge.

MacNeil, H. (1991). Defining the limits of freedom of inquiry: The ethics of disclosing personal information held in government archives. *Archivaria, 32*.

Mayer, T., & Steneck, N. (Eds.). (2012). *Promoting research integrity in a global environment*. Singapore: World Scientific.

Maynard, S. (2017). Review of Katherine Biber and Trish Luker, eds., Evidence and the archive: Ethics, aesthetics and emotion. *Archivaria, 84*, 161–164.

McCullagh, C. B. (2000). Bias in historical description, interpretation, and explanation. *History and Theory, 39*(1), 39–66.

McGregor, H. E. (2015). North of 60: Some methodological considerations for educational historians. *Historical Studies in Education, 27*(1), 121–130.

Nelson, B. (2009). Data sharing: Empty archives. *Nature News, 461*(7261), 160–163.

Nickels, S., Shirley, J., & Laidler, G. (2006). *Negotiating research relationships with Inuit communities: A guide for researchers*. Ottawa: Inuit Tapiriit Kanatami, Nunavut Research Institute.

Nielsen, J. A. (2013). Delusions about evidence: On why scientific evidence should not be the main concern in socioscientific decision making. *Canadian Journal of Science, Mathematics and Technology Education, 13*(4), 373–385. https://doi.org/10.1080/14926156.2013.845323.

NSERC, & SSHRC. (2014). *Tri-council policy statement: Ethical conduct for research involving humans*. Ottawa: Interagency Secretariat on Research Ethics.

Panel on Responsible Conduct of Research. (2016). *Tri-agency framework: Responsible conduct of research*. Ottawa: Government of Canada.

Pegg, J., Wiseman, D., & Brown, C. (2015). Conversations about science education: A retrospective of science education research in CJSMTE. *Canadian Journal of Science, Mathematics and Technology Education, 15*(4), 364–386. https://doi.org/10.1080/14926156.2015.1093202.

Pimple, K. D. (2007). Using case studies in teaching research ethics. *Ethics in Science and Engineering National Clearinghouse, 338*. Retrieved from https://scholarworks.umass.edu/esence/338

Raptis, H. (2010). Dcoumentation as evidence. In A. Mills, G. Durepos, & E. Wiebe (Eds.), *Encyclopedia of case study research*. Thousand Oaks: Sage.

Ritchie, S. M. (2008). The next phase in scholarship and innovative research in science education. *Research in Science Education, 38*(1), 1–2.

Schoenherr, J. R., & Williams-Jones, B. (2016). Early contributions to the evolution of the Canadian scientific integrity system: Institutional and governmental interaction in the policy diffusion process. *Canadian Journal of Higher Education, 46*(1), 57–75.

Schrag, Z. M. (2010). *Ethical imperialism: Institutional review boards and the social sciences, 1965–2009*. Baltimore: Johns Hopkins University Press.

Scott, J. L. (2007). *Exploring the use of virtual field trips with elementary school teachers: A collaborative action research approach*. PhD Thesis, University of Toronto. Retrieved from http://link.library.utoronto.ca/eir/EIRdetail.cfm?Resources_ID=742081&T=F

Simon, J. (2013). Cross-national and comparative history of science education: An introduction. *Science & Education, 22*(4), 763–768. https://doi.org/10.1007/s11191-013-9576-7.

SSHRC Ethics Special Working Committee. (2008). *Extending the spectrum: The TCPS and ethical issues in internet-based research*. Ottawa: Tri-Council.

Steneck, N., Mayer, T., Anderson, M., & Kleinert, S. (2015). *Integrity in the global research arena*. Singapore: World Scientific Publishing Company.

Tinkler, P. (2013). *Using photographs in social and historical research*. Newbury Park: Sage.

Tippett, C. D., & Milford, T. M. (2019). *Science education in Canada: Consistencies, commonalities, and distinctions*. New York: Springer International Publishing.

University of Toronto Libraries. (2019). *Ontario Historical Education Collections (OHEC)*. Retrieved from https://oise.library.utoronto.ca/research-ontario-historical-education

van den Hoonaard, W. C. (2011). *The seduction of ethics: Transforming the social sciences*. Toronto: University of Toronto Press.

van den Hoonaard, W. C., & Hamilton, A. (2016). *The ethics rupture: Exploring alternatives to formal research-ethics review*. Toronto: University of Toronto Press.

van den Hoonaard, W. C., & Tolich, M. (2014). The New Brunswick Declaration of research eth-
    ics: A simple and radical perspective. *Canadian Journal of Sociology, 39*(1), 87–98.
Vervaart, P. (2014). Ethics in online publications. *EJIFCC, 25*(3), 244–251. Retrieved from
    https://www.ncbi.nlm.nih.gov/pubmed/27683471,   https://www.ncbi.nlm.nih.gov/pmc/
    PMC4975197/.
Vogrinc, J., Jurišević, M., & Devetak, I. (2010). *Ethical aspects in science education research.*
    Paper presented at the XIV international organization for science and technology symposium,
    Bled, Slovenia.
Vogrinc, J., Jurišević, M., & Devetak, I. (2019). *Ethical aspects in science education research.*
    Paper presented at the XIV IOSTE symposium, Bled, Slovenia.
Zeller, S. (2000). Roads not taken: Victorian science, technical education, and Canadian schools,
    1844–1913. *Historical Studies in Education/Revue d'histoire de l'éducation,* 1–28.
Zeller, S. (2017). Context, connections and culture: The history of science in Canada as a field
    of study. In *The romance of science: Essays in honour of Trevor H. Levere* (pp. 277–299).
    New York: Springer.
Zimmerman, S. (2015). Responsible conduct of research: A Canadian approach. In N. Steneck,
    T. Mayer, M. Anderson, & S. Kleinert (Eds.), *Integrity in the global research arena*. Singapore:
    World Scientific Publishing Company.

John Allison is a Full Professor of Education in the Schulich School of Education at Nipissing
University. He is a historian of education and did his EdD at the University of Toronto in 1999. His
current primary research projects focus on the history of education diplomacy and comparative
histories of special education. He has written on the topics of the history of technical education in
Canada, research ethics, education diplomacy, and comparative histories of special education. He
has published in peer-reviewed journals including History of Education, the British Journal of
Educational Studies (BJES), the Journal of Educational Administration and History, and Diplomacy
and Statecraft among others. His first book on the history of Canadian education diplomacy, A
Most Canadian Odyssey: Education Diplomacy and Federalism, 1844–1984 was published in
2016 with the Althouse Press, at Western University in London, Canada. In 2014, Dr. Allison was
the recipient of the Chancellor's Award for Excellence in Teaching at Nipissing University.

Michaela Vogt did her PhD at Würzburg University in 2014, afterwards worked at the Universities
of Freiburg and Ludwigsburg as substitute and associate professor and now is employed as full
professor at Bielefeld University where she works since 2018. Her main focus in research is the
theory and history of inclusive pedagogy and education in general. Here she is amongst other
research activities the leader of two funded projects about the history of special needs assessment
procedures in Germany and Canada and about inclusive learning settings in Sweden, Luxemburg,
Germany and Italy with emphasis on teaching materials being used in these settings. Her studies
are mainly based on an international comparative approach as well as on the goal to theorize the
finding and integrate them into a larger theory of inclusion.

Chapter 4
Science Education Practices: Analysing Values and Knowledge

Gerd Johansen and Trine Anker

4.1 Introduction

In this chapter, we present and apply a tool for analysing knowledge and values in science education practices, and we discuss this tool's affordances from a research ethics perspective. In developing the tool, we use Schatzki's (1996) practice theory as a starting point. From this point of departure, science education research and science education are seen as two different practices. The researchers are embedded in the practices of doing research, while the science education practices are the object of research. Both research and science education practices involve different aspects of knowledge and values. However, here, we focus on researchers' investigation and analysis of knowledge and values in science education practices. While knowledge is a frequent theme in research on school science, values are often treated more implicitly. We see it as important to highlight the salience of investigating values as part of these practices. School science is value based in the sense that it includes contents and ways of working that are meant to contribute to the students' growth, both personally and as members of society. For more detailed perspectives on this topic, see, for example, Roberts and Bybee (2014) or Carlone (2014). Researchers are not neutral when they investigate practices, and they bring knowledge and values into the practices of research in general (Macfarlane et al. 2014) and the interaction with school science practices in particular (Jenkins 2000). As values are often an embedded and unspoken part of practice (MacIntyre 1985), there is a risk that the values at stake and value judgement may be hidden.

Both authors of this chapter have extensive experience with empirical research in Norwegian classrooms and schools, and both have worked as teachers in secondary

G. Johansen (✉)
Norwegian University of Life Sciences, Ås, Norway
e-mail: gerd.johansen@nmbu.no

T. Anker
Norwegian School of Theology, Religion and Society, Oslo, Norway

© Springer Nature Switzerland AG 2020
K. Otrel-Cass et al. (eds.), *Examining Ethics in Contemporary Science Education Research*, Cultural Studies of Science Education 20,
https://doi.org/10.1007/978-3-030-50921-7_4

and primary schools for several years. We have transitioned from the role of teachers to that of educational researchers, and thus, become aware of the potential conflicts between different values when analysing school science practices. If the values embedded in the practices are not understood by the researchers, the researchers may present research findings from science education practices in ways that are unrecognisable to the teachers and students. For instance, when researchers present a part of a classroom activity, this can emphasise problematic aspects of the activity at the expense of that which teacher and students find valuable.

In the following, we first outline some research ethics issues connected to our focus on researchers' investigation and analysis of educational practices. Then, we describe our theoretical position—practice theory as a starting point for developing the analytical tool. To illustrate our points, we present an example from a science education study to show how the tool could be used. Finally, we return to the broader field of research ethics and discuss how this analytical tool can meet some of the researchers' ethical challenges by helping the researchers become aware of explicit and tacit knowledge and internal and external values in science education practices.

4.2 Research Ethics: Values in Research

There is a growing interest in studying values in research (e.g., see *Handbook of Academic Integrity*, edited by Bretag (2016)). Denzin and Giardina (2016/2007) refer to respect for individuals, beneficence and justice as common values in qualitative research. However, they critique these values as too narrow, and at the same time, too broad. They are too narrow because they do not cover the field of research ethics; however, as concepts, they can be too broad by becoming what Hammersley (2008) calls vague value commitments. For instance, there are different forms of justice (e.g. social, epistemic and judicial), and as Gewirtz and Cribb (2006) point out, these may be incompatible: The notion of 'treating people justly' can imply different things for different people or in different contexts. For a more elaborate problematisation of justice as a value guiding qualitative research, see Hammersley and Traianou (2014).

Researchers make evaluative (i.e. value) judgements at every stage of the research process, and these may have ethical implications. For instance, they make such judgements in deciding what questions to ask, what evidence to collect and how to interpret and disseminate that evidence (Gewirtz and Cribb 2006; Hammersley 2008). There are also ethical issues that arise from how validity and reliability is handled in research (Fendler 2016). Thus, researchers need to deliberate on the choices they make. Even if these kinds of deliberations and judgements are important components of research and research education, however, there are differences in how ethical issues are argued for and approached (Beach and Eriksson 2010).

In our context, Norway, a precondition for research is that ethical guidelines must be followed. There are rules on the international, national and institutional levels that must be met prior to receiving the necessary approval for research projects. These rules are made to ensure data privacy, confidentiality and informed consent.[1] Hammersley and Traianou (2011) claim that these kinds of 'technical' rules are not necessarily conducive to qualitative research as they stem from other research fields. At least, these rules do not capture all the challenges that arise when conducting research (Lincoln and Cannella 2016/2007).

One of the challenges is managing the relationship with participants. On the 'technical side,' this includes informed consent. On the 'analytical side', this can mean involving key stakeholders/participants in the analysis, which is a good principle. However, according to Levinson (2010), this kind of involvement is difficult to carry out as practice-based research has a tendency to 'flow' in unpredicted directions. By nature, analytical interpretations are not static, and different theoretical frameworks may be tested and adjusted in the process. Even when researchers involve participants in the analysis and writing of research, the researchers have the knowledge about possible frameworks and perspectives, as well as article genres and requirements. Hence, there is a risk that the researchers will state the 'problems' and 'solutions' for the participants. This easily leads to unclear expectations and roles that will influence the relationship between researchers and participants (Hemelsoet 2014). In other words, conducting research with participants is subject to difficulties in the relationship between researchers and participants.

One way of overcoming difficulties in this relationship involves providing the participants the opportunity to engage dialogically in the research. Aluwihare-Samaranayake (2012) states that it is important to give the participants a voice in a research project:

> [W]hen interpreting spheres of people's lives and community experiences, it is crucial to adhere through dialogue and critical consciousness and through an inter-subjective lens to the principles of respect, beneficence, nonmaleficence and justice to ensure that the research is enabling for the participant and facilitates humane transformation to achieve empowerment. (p. 76)

Even if we agree with this position in principle, however, it is difficult to carry out. Researchers need a thorough understanding of the practices in which their participants are engaged to be able to employ an inter-subjective lens. In a discussion concerning insider versus outsider positions in research, Bridges (2009) claims the possibility for outsiders (i.e. researchers) to understand the participants and their practices through patience, persistence, hard work, empathy and imagination. At the same time, there are good reasons to resist an outsider understanding as it can be perceived as the 'arrogance of those who claim to understand us on limited acquaintance' (Bridges 2009, p. 118). Furthermore, sources of resistance to outsider understanding may be the participants' desire to protect their privacy or even fear of a loss of identity or social belonging (Bridges 2009).

---

[1] EU General Data Protection Regulations (GDPR), national Norwegian Centre for Research Data (NSD;   http://www.nsd.uib.no/nsd/english/index.html)   and   National   Research   Ethics Committees (NESH).

Another challenge is the difficulty of predicting how research results may affect broader social relations and the political implications of such effects (Beach and Eriksson 2010). It is problematic to assume the consequences of the analytical process. Hence, taking part in a research project is 'risky business'. Gewirtz and Cribb (2006) argue that, in addition to being aware of values as part of the research process, researchers have responsibility for the (value) implications of their work. This is especially important since there may not be a clear separation between knowledge production and use of research knowledge (e.g. through planned interventions in a practice), which is often the case in educational research. Gewirtz and Cribb (2006) call for an ethical reflexive approach that include the following: acknowledging and responding to tensions between the various values that are embedded in the research, taking the practical judgements and dilemmas of the practices that are researched seriously and taking responsibility for the political and ethical implications of the research. Hammersley (2008), on the other hand, argues against this meshing of knowledge production and use of knowledge because it is too easily influenced by the researchers' values—and thus, implications of research and even research evidence may become biased. Although we are aware, and see the importance of Hammersley's objection, our position is that research in education will often aim for changing a practice. Therefore, we see it as important to have ways of dealing with tensions between different values. This implies that normative judgements and values in practice need to be investigated.

4.3   What Is Practice?

We use the theory of social practices, or practice theory for short (Schatzki 1996), to strengthen the awareness of how knowledge and normative judgements are intertwined in educational practices. While research ethics has tended to focus on the relationships between researchers and research participants, instead, our approach involves focusing on practice, which opens up seeing activities as bodily *and* mental routines (Reckwitz 2002b) that are made possible but also constrained by structures (Schatzki 2001). Furthermore, there is a normative aspect to practices as they are evaluated by those who carry them out (Schatzki 1996). In this section, we discuss how practices can be understood and what this may mean for notions of values and knowledge.

To say that education is a practice is based on a present debate in the philosophy of education, and to a great extent, this view leans on the work of neo-Aristotelian researchers (Dunne and Hogan 2004). They oppose the idea of education as instruction, and they see the life of schools as an important focus of research. In this understanding, knowledge and values are both embedded in practices and important to study.

The concept of practice has been developed in interdisciplinary practice theory to overcome the actor–structure division, common in social science research (Schatzki 2001). Practice theorists will say that neither an overall structure forcing people to act as they do nor a totally open, unguided space for acting freely, exists. Rather, some common norms or expectations define humans' actions and are necessary ingredients of certain practices. Schatzki (2001, p. 11) defines practice as 'embodied, materially mediated arrays of human activity centrally organized around shared practical understanding'. According to Schatzki (2001), activity is looser and less structured than practice is. As such, all practices are based on activities but not all activities are part of practice. Arrays are arrangements or ordered sets of entities, and they point towards practices as social order and patterns of activities. However, Schatzki (1996) states that social order is established within a social practice, meaning that the order does not exist as something outside or beyond the practice. Furthermore, Reckwitz (2002a) emphasises the embodied and material part of social practices. When we engage in a practice, we use the body in certain ways and handle different material objects in our material surroundings. To state that human activity is embodied means that actions are performed by the body and thus, include mental and emotional activities (Reckwitz 2002b).

4.3.1 Practices and Values

Schatzki's (2001) outline of practice overcomes the mind–body dichotomy. He states that a practice includes thoughts, emotions and bodily activity. However, the aspect of values is not elaborated on in Schatzki's (Schatzki 1996, 2001, 2013) practice theory. In his book *After Virtue*, MacIntyre (1985) criticises our society and its heritage from the Enlightenment period for separating values and morals from more mechanical descriptions of human activity. MacIntyre (1985) uses the term 'goods', which is equivalent to our use of 'values'. However, we use 'goods' in this section for consistency. His definition of practice is as follows:

> any coherent and complex form of socially established cooperative human activity through which goods internal to that form of activity are realized in the course of trying to achieve those standards of excellence which are appropriate to, and partially definitive of, that form of activity, with the result that human powers to try to achieve excellence, and human conceptions of the ends and goods involved, are systematically extended. (MacIntyre 1985, p. 187)

MacIntyre (1985) highlights the normative aspects of practices, stating that internal goods are necessary to call something a practice. As a result of focusing on the normative aspects, MacIntyre's (1985) outline of practices has been taken up by philosophers of education (Dunne 2003; Hogan 2003; Noddings 2003; McLaughlin 2003), and his definition of practice has been used to emphasise the value dimensions of educational practices. However, MacIntyre (1985) does not regard education as a practice, because he sees education as teaching and instruction. In contrast, in a thorough discussion on the topic, Dunne and Hogan (2004) disagree with what they consider to be an instrumentalist view of education and conclude that, to them, education *is* a practice.

MacIntyre (1985) emphasises that there is a common understanding of what to strive for in a specific practice. Something can either be good and right or it can be wrong. These standards can be seen as the collective aims of a specific practice (MacIntyre 1985). Taking part in a practice means working with others to achieve goals. MacIntyre (1985) divides the aims of a practice into two main categories—external and internal goods. External goods can be, for example, power, money and fame. These can be goods in many practices and are not specific for one practice. Such goods become the individuals' possessions, and the more one has, the less there is available for other people to obtain. For instance, not everyone partaking in a practice can have the same possibility to gain power. In other words, external goods have limits; however, internal goods are realised when trying to achieve excellence in practice. In contrast to the individualistic aspect of competing for external goods, the achievement of internal goods is a good for the whole practice, and thus, such goods are not delimited to individual participants. Internal goods are specific to a practice, and therefore, they can only be identified and recognised by the experience of participating in the practice in question. Hence, '(t)hose who lack the relevant experience are incompetent (…) as judges of internal goods' (MacIntyre 1985, p. 188). Such a statement may be an argument in favour of taking part in the practices that are being studied to fully understand what is at stake.

To support the idea that education can be called practice, internal goods are essential. An example of an important internal good to strive for in the science classroom could be that all students contribute constructively and jointly towards building knowledge, which will improve the practice. Moreover, there are also external goods in educational practices. National and international tests of science knowledge involve ranking schools or countries, and often it is important to be the 'best' because it increases prestige. However, for most, this is an unachievable aim.

MacIntyre (1985) emphasises the importance of institutions for practices' long-term survival. A practice is not the same as an institution, but it does depend on an institution, while an institution simultaneously depends on practices (MacIntyre 1985). For example, science education research can be seen as several different practices. For some of these research practices, the internal goods are improving science education, as well as striving towards excellence in science education research. However, the institution—the university—must ensure that external goods are achieved; for example, they must make certain that articles are submitted to show academic production and secure necessary funds to survive as an academic institution. Hence, there can be a conflict between the aim of the research practice and the institution bearing this practice (cf. Löfström 2016). The same can be said about the practices of science education in schools. Science education strives for better science education, whereas the school, as an institution, must also strive for various external goods. In the rest of the chapter, we revert to the term 'values' instead of 'goods'.

4.3.2 Practice and Knowledge

There are certainly things going on in a science classroom that have little to do with the subject, for instance, quarrels between students, faulty technical equipment or unexpected changes in the schedule. However, in science classrooms, the main activity is dealing with science knowledge. Researchers need to be aware of how knowledge in science education practices plays out—how knowledge is approached, developed and shared among the participants in a given setting (Jensen et al. 2015). This allows the researchers to move beyond what Jensen et al. (2015) call the traditional emphasis on 'knowledge as content' to knowledge as investigative processes, modes of inquiry and principles for verification as components of school activities. This means that there is no strict division between knowledge and skills. In practice as performance (Reckwitz 2012), i.e. when a practice is carried out, knowledge is one aspect of embodied human activity; knowledge can be part of what people think, say and do. In practice theory, material objects can be part of how knowledge is developed and shared. Extending this argument somewhat, a focus on practices allows for the incorporation of tacit forms of knowledge (Collins 2001).

Collins (2010) discusses explicit and tacit knowledge, making an elaborate argument that there are different forms of tacit knowledge, namely, weak, medium and strong tacit knowledge. Weak tacit knowledge passes between participants in a practice when they have enough cultural similarity. It is possible to make this type of tacit knowledge explicit with some effort via, for example, longer and more substantial explanations or apprenticeships (e.g. learning some skill). However, making the tacit explicit may be 'impossible' because this is cumbersome within a reasonable timeframe and the limits of the human attention span. Furthermore, a participant in a practice may not see that some of the knowledge he or she applies is important; the knowledge is not recognised, and thus, it is left uncommunicated. It is also possible that he or she does not see that the knowledge is needed by the other participants, so it is left unsaid. Medium tacit knowledge involves how people perceive and use the body in performing knowledge (i.e. skills). Strong tacit knowledge is collective at the societal level. Individuals share collective social knowledge by partaking in social practices, for example, how to walk on a sidewalk when it is or is not crowded. This view of knowledge has consequences for education: '[E]ducation is more a matter of socialization into the tacit ways of thinking and doing than transferring explicit information or instructions' (Collins 2010, p. 87).

Knowledge inherent in a practice builds on previous practice and includes a historical or traditional aspect of practice. However, some practices, such as teaching and research, are also focused on developing, applying and sharing new knowledge:

> [I]t is also a characteristic of current times that many occupations and organizations have a significant knowledge base. In these areas, one would expect practitioners to have to keep learning, and the specialists who develop the knowledge base to continually reinvent their own practices of acquiring knowledge. (Knorr Cetina 2001, p. 175)

Knorr Cetina (2001) continues by saying that, when a practice aims at developing knowledge, the practice becomes something other than just habitual activities. One way of developing knowledge in an educational practice is making the aim to improve the practice. According to Turner (2001), the learning of, and thus participating in, some practices purpose relative. We assume that practices like teaching and research are purpose driven in that they seek to improve practices. This implies that these practices will rely on an amalgam of values and knowledge.

To summarise, we see science education and research on science education as practices with inherent aspects of knowledge, both tacit and explicit, as well as internal and external values. These different aspects are intertwined. However, in the next section, we employ an analytical split between values and knowledge in practices for the sake of performing the analyses.

4.4 An Analytical Tool: Practice—Knowledge and Values

So far, we have examined how practices can be understood in terms of values and knowledge. As we have seen, practice theory implicates a broad view on knowledge. A practice's inherent knowledge is both tacit and explicit (explicitly communicated), where tacit knowledge is more elusive because it is not expressed through verbalised language. Besides these different forms of knowledge, values are important aspects of a practice. Values provide participants with reason and direction when working with knowledge (striving for knowledge). In the differentiation between external and internal values, internal values of a practice are the 'real' values, but they are seldom verbalised. While internal values are specific to a practice, external values are common to different practices. Values seem to be quite fixed; MacIntyre (1985) claims, 'In the realm of practices the authority of both values and standards operates in such a way as to rule out all subjectivist and emotivist analyses of judgment' (p. 190). There is a normative aspect of practice that is shared among the practitioners for joint judgement. However, values can be criticised—a not uncommon feature of classroom practice.

Knowledge and values are intertwined and operate on different levels of verbalisation and visibility in a practice. To make it possible for the researchers to differentiate between these different levels, we have developed a tool for the analytical process (see Fig. 4.1). Activities that constitute the practice can be interpreted according to the four categories given in the figure. We emphasise that this

Tacit knowledge & internal values	Explicit knowledge & internal values
Tacit knowledge & external values	Explicit knowledge & external values

Fig. 4.1 Practice as configurations of knowledge and values

analytical tool is made for analytical purposes, so it does not reflect what *is*, but rather, helps in understanding the practice.

For the teacher (or researcher), there may not be a perceived conflict between internal and external values. Different positions on knowledge and values may exist side by side in a practice. The teacher may be explicit on some of the purposes, knowledge and values while not stating others as clearly. In the next section, we provide an example to show how the tool can be used to analyse knowledge and values in educational practices. The main function of the example is generating a starting point for investigating how knowledge and values are intertwined.

4.4.1 A School Science Research Project: An Example for Analysis

Before applying the analytical tool, we want to provide an overview of a research project, as well as an example story. This story is based on various fieldwork notes and interviews from one of the authors, written into one coherent narrative. It is written in first person to gain the researcher's perspective on the fieldwork: This is a story based on the researcher's gathered facts, as well as her thoughts, feelings and actions. A narrative can be a means through which one attempts to grasp the real; it also facilitates the reader's engagement with this particular reality (Watson 2011). The reason for writing a narrative is to synthesise years of fieldwork and include several significant incidents. In this way, we are able to form these incidents into a single story, which in turn, can be analysed in the frame of an article. According to Connelly and Clandinin (1990), the criteria of verisimilitude and appearance trump reliability and validity criteria when presenting research findings through narratives. The truth of the story's timeline is not that important; what is more important is to show how the discussion can be transferrable and useful in other research projects (Connelly and Clandinin 1990).

The example story is taken from an ethnographic project in a science education class where I (first author) spent a great deal of time with the teacher and students. The aim of the research project was to support students' meaning making when engaged in structured inquiries. The teacher facilitated the activities, and the students made and appropriated a broad spectrum of representations, including some traditional (e.g. tables and graphs) and other less traditional representations (e.g. cartoons and photo stories) in school science. The empirical material was analysed using discourse analysis inspired by Halliday and Matthiessen (2004) and Fairclough (2003).

The teacher was experienced, and she expressed an affinity for working with students. She wanted to contribute to their education and 'make a difference'. The students, aged 16 years, were mostly low to medium achievers in terms of the subject matter. They wanted to do well in school and pass their final exam in science that spring, but they did not have a special interest in school science. Most of the time, the students complied with the teacher's instructions, and they worked together quite amicably.

The following fieldnotes, with questions/comments in italics, are from a lesson where students were instructed to perform two small practical activities in thermodynamics, on the topic of heat:

> The students are stalling in their practical activities. 'What are we to do now?', they ask. Some are doing something else entirely. *They did not 'understand' the teacher's two previous introductions. The question is why? Is the topic of heat not very engaging? Do they understand why they are doing this?* The teacher 'blows the whistle', and for the third time, introduces the practical activities. She provides a detailed demonstration of how to use the equipment. Without being impatient, she asks the students, 'What do you do then?' One or two raise their hands to answer. At the very end, the teacher asks, 'Ok?'; some students say 'yes', some nod and some do not give a visible/audible response. The teacher is very calm. The teacher focuses on what to do; all the verbs she uses are physical actions to carry out the steps in the procedure. There are no verbs that indicate that the students are to try to make scientific observations.

In my view, the students had to interpret the practical activities independently, unsupported by the teacher. There was no emphasis on how to make observations (e.g. 'If you touch the beaker, what do you feel?' or, 'Look at the bottom …') and inferences (e.g., 'What do you think this means?'). In my original analyses of the fieldwork, I found it problematic that the teacher's practice focused on 'doing' and not on 'meaning'. In other words, the goal of *doing science* was upheld at the expense of the other goals, which has been seen as a persistent problem in science education (Hodson 1993; Windschitl 2008; Gyllenpalm et al. 2009). The teacher did little, in my view, to use *doing science* as a starting point to engage students in a discussion on *learning about science* (see e.g. Hodson 2014) and broadening their conceptual understanding of heat. At a personal level, I had (and have) a deep respect for this teacher. However, I felt I had to report what I found problematic in the practice, and I evaluated it as science education with 'some problems'. I pointed to the teacher's somewhat narrow understanding of the purposes and knowledge concerning students' practical activities. The teacher told me that she incorporated practical activities when she thought it was important for visualising some phenomenon or concept, and to her, it was important that the activities were fun and easy to carry out for the students. The teacher sometimes 'complained' about the voluminous and extremely specified national science curriculum, which left her with little possibility to manoeuvre the science subject for students with varying backgrounds in science and little experience with laboratory work.

As these are persistent problems that are frequently reported in science education literature, I may have had a special critical gaze when considering the classroom activities because of my academic training. The problems became visible through the analysis of the verbal communication. However, afterwards, I felt that I had missed something in my account. My hunch was that I had missed some of the unspoken workings of this practice.

4.4.2   Applying the Analytical Tool to the Example

As we discussed above, researchers make value judgements at every stage of a research process. Even if the aim of using the tool is to open up knowledge and values in practices, researchers' use of the tool can never be value neutral. In the example, the researcher judged the teaching and learning situation against her standards of what good teaching practice in science entails. These standards are related to how teachers and students approach, develop and share knowledge. The researcher has developed these standards by reading science education literature, as well as gleaning them from personal experiential context. Clearly, the teacher had standards that were more connected to the context of which she was a part, and her standards were not limited to notions of knowledge. In other words, the valuing processes for both teacher and researcher were strongly connected to their respective experiences. Value judgements often start with what can be regarded as a gut feeling: 'This is wrong', or 'This is good'. To translate this into a more formal language, the researcher used a vocabulary that was taken from the literature. The researcher observed the classroom activities and looked for certain cues (e.g. type of verbs) when judging their quality. Engagement in understanding and judging practice draws on knowledge, emotions and expectations, and it is largely implicit and historically–culturally specific (Reckwitz 2002b). The researcher was familiar with these types of translations, especially during the analytical process. This was part of her developed research skills, where the researcher often retrospectively analysed these first intuitive value judgements—and possibly, 'corrected' them. The teacher, in contrast, had to act in the classroom with minimal time to think. While teachers act in the classroom, education researchers describe and analyse their acts. We now broaden the description by accounting for values and knowledge.

The first step on the way to making an analytical description is identifying the explicit and tacit knowledge in the activity. There were several different types of explicit knowledge in the teacher's presentation. She used words like 'heat' and 'heat transfer' when she talked to the students and thus, directed the attention to central thermodynamic entities. By showing and telling, she connected these entities to the use of the equipment. There were also several hints of tacit knowledge in this short presentation. The teacher never explained (to the researcher's knowledge) the importance of following a strict procedure to the students, and thus, she did not clarify the need for students to remember the procedural steps and their sequence. There was no reference to making observations or inferences from observations. That these observations could, and perhaps ought to, be interpreted in terms of general principles was left tacit or implicit. How the entities were to be linked to the students' experiences was not mentioned here, although it was later touched on, at least to some degree. The teacher did not provide reasons or purposes for doing the activity; therefore, it was given no explicit value.

The next step is identifying possible values that are embedded in the practice. When the teacher and researcher spoke about what they saw as valuable in (science) education, there seemed to be meaning alignment. On the surface, the values were shared; however, the meaning alignment cannot be taken for granted because values are seldom made explicit and clarified. The researcher's interpretation of values relied on her ethnographic field experience, which is more extensive than the sum of her fieldnotes, interviews and videos (Hammersley and Atkinson 2007). Some of the external values in this practice were the explicit notions from the national government and school leadership to raise standards, and specifically, increase the rate of passing grades. This can be characterised as an external value for the school as an institution (i.e. achieve higher grades, be a 'better' school and attract more academically gifted students), although it has a 'twin' internal value: The teacher wanted the students in the class to do well. For the teacher, an important aspect of 'doing well' meant students could master what she saw as the standards of school science. The school administrators' anticipation that more students would pass, was problematised by the teacher and researcher from a 'raise the standards of science education' point of view. The external value (passing) created an incentive for the teacher to 'lower standards' in the sense that she had to compromise on what she thought was valuable. The internal values of this practice were about striving to adopt and adapt to a (school) science way of thinking and acting. What was seen as valuable was remembering factual knowledge and being accurate when doing practical work.

We now use the analytical tool to delve deeper into the example, which provides an opportunity to examine different combinations of explicit and tacit knowledge, as well as internal and external values.

Explicit Knowledge and Internal Values The teacher chose words like 'heat' and 'heat transfer'. She could have chosen other terms, such as 'energy transfer'. Choosing more everyday words can be seen as contrary to the internal value of adopting the 'science way of thinking'. By making this choice, she communicated in a language that was closer to the students' spoken language. Maybe because she wanted them to feel included by not using an alienating language. When the teacher went through the procedure (for the third time), she emphasised the value of following procedures and how to use equipment in science. She wanted the students to remember what to do. The value of remembering could be coupled with the teacher's conviction that all students should be included and able to do the activity. Thus, the students could learn the necessary scientific knowledge so they could adopt vital parts of (school) science. However, as science is a subject where there is a considerable amount of information to remember, the teacher selected the knowledge (within the limitations of the national curriculum) on which she placed particular value.

Explicit Knowledge and External Values In this example, verbalised knowledge had another characteristic: It prepared students for the final exams. This was an external value for both the teacher and students. It was necessary to prepare students for their exam throughout the year, even when it was not explicitly communicated that this was the focus. It was important for the students to achieve good results, and the teacher wanted her students to perform well. Even if seldom explicated, 'good teachers' manage to teach their students so that they achieve 'good results'. In some ways, students achieving good results elevates a teacher's status. For the school, good grades are important because they may be used to indicate to school authorities that this is a school that did achieve 'the results'.

Tacit Knowledge and Internal Values The teacher gave no specific reasons either to her students or the researcher about why it was so important to follow the set procedure. It just *was* important. The procedure became a way of tacitly adapting to a (school) science way of thinking and acting. The procedure needed to be followed, but its inner logic was never explained. Moreover, observations and inferences were not really touched on by the teacher, and they were probably seen as aspects that were 'just going to happen'. Perhaps observations became tacit since they were performed bodily and obvious for the teacher (cf. Collins (2010)). The teacher had a strong motivation to let her students experience and feel curiosity and enjoyment when doing practical work. If she were to focus on observations and inferences, some of the valuable curiosity could vanish. This activity had similarities to discovery learning where the results are waiting at the end of the procedure, and these results will be incontestable, see Gyllenpalm et al. (2010) for an elaboration on different traditions in practical work. This activity can be seen as part of the enculturing of students into tacit ways of thinking and doing school science. The internal values, a mixture of wanting the students to be able to carry out the procedure accurately and desiring that they feel curiosity and enjoyment of science, constituted a science subject where knowledge will appear when the procedure is followed correctly. Science became only loosely coupled to general principles: The phenomenon was not seen *as* something.

Tacit Knowledge and External Values There is another way to understand the values connected to tacit knowledge: If the aim is to teach students so that they pass the final exam, there is no real need to delve into how science works or the importance of observation and inference. In this case, observations and inferences could be left tacit because they were less needed on the exam. As the teacher once said during an interview, '*If the students were able to recall detailed knowledge correctly, the external censor[2] will be impressed*'. This is what the average censor would expect, she claimed. By the students' accurate recitation of core knowledge, the teacher— and students—would probably be considered 'successful'.

To summarise, in this practice, explicit and tacit knowledge and internal and external values were all at play. The practice of science education is a complex amalgam of knowledge and values, which can create problems for a researcher. If a researcher only 'sees' the 'visible' practice and omits values and the un-verbalised, there is a risk that the analysis of the school practice will be too shallow.

We used the tool to explore configurations of knowledge and values. This allowed us to go beyond the verbalised knowledge to study embodied and tacit knowledge.

---

[2] In Norway, the final exams have an external censor. The censor is a teacher from another school that assesses the students' performance. If a teacher chooses to do something unconventional in science, this may not be well received by some of the censors.

Internal and external values gave the knowledge in this practice worth. Values contributed to increasing the depth in the understanding of the practice compared with focusing only on verbal communication.

The analytical process had three main stages, each providing the researcher with different forms of insight into the practice. The main stages were as follows:

1. The researcher's initial gut feeling: The lack of meaning and purpose concerning knowledge and ways of working in the activity;
2. Strengthening of the gut feeling through the original analysis of speech acts: No reason was given for why the students should follow the procedure—or indeed, why it was important to do this at all. The analysis also revealed a lack of verbs that could have directed students towards observations and inferences, connecting the concepts in the teacher's introduction with the activity; and
3. Elaborating on the original analysis by applying this analytical tool, considering the values of the practice: In the interplay and conflict between external and internal values (e.g. more students with passing grades and a 'science way of thinking'), the approach to the science knowledge became less self-evident. What should a teacher do to embrace very different values? At the same time, the choices made by the teacher in the practice becomes more complex. Hence, it is easier to argue that the teacher's sayings and doings are highly reasonable.

For the third stage, we open up the different forms of knowledge and values that are intertwined in any practice.

4.5 Ethical Challenges When Analysing School Science Practices

The 'intertwined-ness' of knowledge and values plays a vital part in science education. In science education, there are debates—and sometimes disagreements—about what the most valuable knowledge is and what knowledge students ought to learn. This can be reformulated into visions for what it means to be knowledgeable in school science (see e.g., Roberts 1988, 2011; Roberts and Bybee 2014; Liu 2013). Moreover, the vision(s) for science education can be seen as the goal of tacit and explicit socialisation into the knowledge culture of (school) science. When researchers and practitioners come together, they do not necessarily share the same visions for what it means to 'be knowledgeable', and they ascribe different values to different aspects of the school subject. For instance, for researchers, the value of incorporating aspects of nature of science explicitly when students do practical activities is often seen as a highly important part of being knowledgeable; see Lederman and Lederman (2014) for an elaboration. In contrast, for teachers, being knowledgeable needs to be negotiated with how they perceive their students' interest and their interpretation of the curriculum. Hence, it makes sense to untangle knowledge and values through an analytical process; otherwise there is a risk that the analysis gives rise to misunderstandings between practitioners and researchers. This position has some ethical consequences for how to approach research.

We are aligned with Gewirtz and Cribb (2006, 2008) and Hammersley (2008) in the view that there is a need to go beyond the technical approaches to research ethics and look into ethical aspects at all stages of the research. The analytical tool makes it possible to identify and acknowledge value tensions and dilemmas that are embedded in practices (Gewirtz and Cribb 2006). When interpretations of practice do not consider what is regarded as valuable in a particular practice, practitioners may not feel that they have been understood (Bridges 2009), and there is a risk that the relationships between researchers and practitioners will become strained. Values are seldom made explicit, and thus, they may be underlying other conflicting issues that cause strain. This strain on the relationships has been discussed in the research ethics literature (e.g. Aluwihare-Samaranayake 2012). However, we do not see the solution to this kind of problem as solely involving improving the relationships between practitioners and researchers. As Bridges (2009) describes, it is possible for researchers to understand a practice through patience, persistence, hard work, empathy and imagination. However, we would emphasise the need for consciously investigating the values that embed the practice, since values are important 'drivers' for choices and actions. For the researchers, to discuss the values in a practice with the practitioners can be one way of validating what is seen as worthwhile, as well as gaining greater reflexivity for all parties. By connecting values to other key elements (in our case, knowledge) one can analytically unravel a complex practice—at least in part.

Hammersley (2008) claims that there is a risk of research bias, that is, for the researchers to produce data and interpret them in ways that are in line with their commitments or prior assumptions. This risk is especially prominent when a researcher's aim is to change or improve practice. As research can never claim to be neutral, it is easy to try to impose one's values when interpreting the practice (Hemelsoet 2014). One way of reducing the risk of bias is through making not only the practices' internal and external values, but also the researchers' values, explicit. The researcher's awareness of own values can mitigate some of the effect of the unpredictability of the analytical process (Levinson 2010) by laying open the communication on important values.

Another problem Hammersley (2008) points out is the problem of scientism in research that aims at changing practices. As he describes it, scientism is a rather narrow understanding of research results and how they can be used to make improvements and affect a practice. Research, he states, can obviously influence a practice, but this should be for the practitioners to decide (Hammersley 2008). We agree that this may be an ethical problem: Can 'we' know what the best changes for a particular practice are? If the intertwined values and knowledge in the practices are not untangled—and the practices' internal values are not acknowledged—there is, as we see it, a great risk that practice will not be improved by the research.

Hammersley and Traianou (2011) argue that the main objective for research is to produce 'sound knowledge', where ethical reasoning is one part of the research process. They warn against approaches to research ethics where values that are external to the task of producing 'sound knowledge' are treated as if they were central to it. Moreover, in a review on research publications in social science, Löfström (2016) discusses several dangers coupled to publication pressure. This pressure may lead researchers to 'overreport' results, minimise research context and downplay the role of their own values. The researchers' assumptions about social interaction and knowledge production influence the choices of theory, research questions, research methods and avenues for disseminating results. These assumptions involve 'personal and social values that can have moral consequences through the choices and actions that researchers take' (Payne 2000, p. 308). This calls for ways to identify and deliberate on the values that are central to the stakeholders in research projects. We propose this analytical tool as a way of mapping and describing knowledge and values in research practices as well.

4.6 Concluding Remarks

Within practices, there are common norms and expectations for what is said and done (Schatzki 1996). This implies that the aims of a practice give value to what is said and done: The combination of explicit and tacit knowledge is intertwined with what is regarded as valuable. Values are a ubiquitous part of any practice and shape what occurs. By differentiating values into external and internal values (MacIntyre 1985), the intertwined knowledge and values can be made visible. We see it as important to explicitly identify what is—and what is not—valuable. This requires that the views of research as free from value judgements are regarded as 'illusions'; however, it does not mean that 'anything goes': Values need to be made clearer as part of the analytical process. We claim that the entanglement of underlying values and knowledge in science education practice needs to be considered during the analytical process.

We have argued that investigating the configuration of values and knowledge as part of the analytical process can be an element in ethically sound research. More precisely, to conduct 'good research', transforming vague and elusive parts of an educational practice into substantial argumentation is important. Moreover, by emphasising both internal and external values, in addition to tacit and explicit knowledge, the tensions and dilemmas the participants are facing become clearer, and the practical judgements within the researched practices can more easily be understood. Hence, the analytical tool presented in Fig. 4.1, can help researchers become ethically reflexive during the analytical process.

There is a need to explore the usefulness of the analytical tool in other school (science) practices. One possible extension is to emphasise materiality and affective aspects in the practices. Such perspectives are important in education, and they are seen as significant in practice theory (Reckwitz 2012). Another possibility is using the tool as a starting point for exploring political consequences of worthwhile knowledge, and by so doing, avoiding the critique of practice theory stating that it does not put enough emphasis on power and micro-politics (Sayer 2013).

Research and education are two practices with different internal values. The tool presented in Fig. 4.1, can contribute making values in different practices explicit. This is beneficial when the goals of educational research are to critically investigate, and possibly improve, the practice of education while at the same time work for its own internal values. Therefore, opting for ethically sound research means incorporating values into the analytical process as a part of the practice.

References

Aluwihare-Samaranayake, D. (2012). Ethics in qualitative research: A view of the participants' and researchers' world from a critical standpoint. *International Journal of Qualitative Methods, 11*(2), 64–81. https://doi.org/10.1177/160940691201100208.

Beach, D., & Eriksson, A. (2010). The relationship between ethical positions and methodological approaches: A Scandinavian perspective. *Ethnography and Education, 5*(2), 129–142. https://doi.org/10.1080/17457823.2010.493393.

Bretag, T. (2016). *Handbook of academic integrity*. Singapore: Springer.

Bridges, D. (2009). Education and the possibility of outsider understanding. *Ethics and Education, 4*(2), 105–123. https://doi.org/10.1080/17449640903326714.

Carlone, H. B. (2014). Cultural perspectives in science education. In N. G. Lederman & S. K. Abell (Eds.), *Handbook of research on science education* (pp. 651–670). London: Routledge.

Collins, H. M. (2001). What is tacit knowledge? In T. Schatzki, K. K. Cetina, & E. Von Savigny (Eds.), *The practice turn in contemporary theory* (pp. 107–119). Oxon: Routledge.

Collins, H. M. (2010). *Tacit and explicit knowledge*. Chicago: University of Chicago Press.

Connelly, F. M., & Jean Clandinin, D. (1990). Stories of experience and narrative inquiry. *Educational Researcher, 19*(5), 2–14.

Denzin, N. K., & Giardina, M. D. (2016/2007). Introduction. Ethical futures in qualitative research. In N. K. Denzin & M. D. Giardina (Eds.), *Ethical futures in qualitative research. Decolonizing the politics of knowledge* (pp. 9–43). Oxon: Routledge.

Dunne, J. (2003). Arguing for teaching as a practice: A reply to Alasdair MacIntyre. *Journal of Philosophy of Education, 37*(3), 353–370. https://doi.org/10.1111/1467-9752.00331.

Dunne, J., & Hogan, P. (2004). *Education and practice: Upholding the integrity of teaching and learning*. Wiley.

Fairclough, N. (2003). *Analysing discourse. Textual analysis for social research*. Oxon: Routledge.

Fendler, L. (2016). Ethical implications of validity-vs.-reliability trade-offs in educational research. *Ethics and Education, 11*(2), 214–229. https://doi.org/10.1080/17449642.2016.1179837.

Gewirtz, S., & Cribb, A. (2006). What to do about values in social research: The case for ethical reflexivity in the sociology of education. *British Journal of Sociology of Education, 27*(2), 141–155. https://doi.org/10.1080/01425690600556081.

Gewirtz, S., & Cribb, A. (2008). Differing to agree: A reply to Hammersley and Abraham. *British Journal of Sociology of Education, 29*(5), 559–562. https://doi.org/10.1080/01425690802381577.

Gyllenpalm, J., Wickman, P.-O., & Holmgren, S.-O. (2009). Teachers' language on scientific inquiry: Methods of teaching or methods of inquiry? *International Journal of Science Education, 32*(9), 1151–1172. https://doi.org/10.1080/09500690902977457.

Gyllenpalm, J., Wickman, P.-O., & Holmgren, S.-O. (2010). Secondary science teachers' selective traditions and examples of inquiry-oriented approaches. *NorDiNa, 6*(1), 44–60.

Halliday, M. A. K., & Matthiessen, C. M. I. M. (2004). *An introduction to functional grammar*. London: Arnold.

Hammersley, M. (2008). Reflexivity for what? A response to Gewirtz and Cribb on the role of values in the sociology of education. *British Journal of Sociology of Education, 29*(5), 549–558. https://doi.org/10.1080/01425690802263718.

Hammersley, M., & Atkinson, P. (2007). *Ethnography: Principles in practice.* London: Routledge.

Hammersley, M., & Traianou, A. (2011). Moralism and research ethics: A Machiavellian perspective. *International Journal of Social Research Methodology, 14*(5), 379–390. https://doi.org/10.1080/13645579.2011.562412.

Hammersley, M., & Traianou, A. (2014). An alternative ethics? Justice and care as guiding principles for qualitative research. *Sociological Research Online, 19*(3), 1–14. https://doi.org/10.5153/sro.3466.

Hemelsoet, E. (2014). Positioning the educational researcher through reflections on an autoethnographical account: On the edge of scientific research, political action and personal engagement. *Ethics and Education, 9*(2), 220–233. https://doi.org/10.1080/17449642.2014.925033.

Hodson, D. (1993). Re-thinking old ways: Toward a more critical approach to practical work in school science. *Studies in Science Education, 22*, 85–142. https://doi.org/10.1080/03057269308560022.

Hodson, D. (2014). Learning science, learning about science, doing science: Different goals demand different learning methods. *International Journal of Science Education, 36*(15), 2534–2553. https://doi.org/10.1080/09500693.2014.899722.

Hogan, P. (2003). Teaching and learning as a way of life. *Journal of Philosophy of Education, 37*(3), 207–224. https://doi.org/10.1111/1467-9752.00321.

Jenkins, E. (2000). Research in science education: Time for a health check? *Studies in Science Education, 35*(1), 1–25. https://doi.org/10.1080/03057260008560153.

Jensen, K., Nerland, M., & Enqvist-Jensen, C. (2015). Enrolment of newcomers in expert cultures: An analysis of epistemic practices in a legal education introductory course. *Higher Education, 70*(5), 867–880. https://doi.org/10.1007/s10734-015-9872-z.

Knorr Cetina, K. (2001). Objectual practice. In T. Schatzki, K. K. Cetina, & E. von Savigny (Eds.), *The practice turn in contemporary theory* (pp. 184–197). London: Routledge.

Lederman, N. G., & Lederman, J. S. (2014). Research on teaching and learning of nature of science. In N. G. Lederman & S. K. Abell (Eds.), *Handbook of research on science education* (pp. 600–620). New York: Routledge.

Levinson, M. P. (2010). Accountability to research participants: Unresolved dilemmas and unravelling ethics. *Ethnography and Education, 5*(2), 193–207. https://doi.org/10.1080/17457823.2010.493407.

Lincoln, Y. S., & Cannella, G. S. (2016/2007). Ethics and the broader rethinking/reconceptualization of research as construct. In N. K. Denzin & M. D. Giardina (Eds.), *Ethical futures in qualitative research. Decolonizing the politics of knowledge.* London: Routledge.

Liu, X. (2013). Expanding notions of scientific literacy: A reconceptualization of aims of science education in the knowledge society. In N. Mansour & R. Wegerif (Eds.), *Science education for diversity. Theory and practice* (pp. 23–39). Dordrecht: Springer.

Löfström, E. (2016). Academic integrity in social sciences. In T. Bretag (Ed.), *Handbook of academic integrity* (pp. 714–728). Singapore: Springer.

Macfarlane, B., Zhang, J., & Pun, A. (2014). Academic integrity: A review of the literature. *Studies in Higher Education, 39*(2), 339–358. https://doi.org/10.1080/03075079.2012.709495.

MacIntyre, A. (1985). *After Virtue.* Notre Dame: University of Notre Dame Press.

McLaughlin, T. H. (2003). Teaching as a practice and a community of practice: The limits of commonality and the demands of diversity. *Journal of Philosophy of Education, 37*(3), 339–352. https://doi.org/10.1111/1467-9752.00330.

Noddings, N. (2003). Is teaching a practice? *Journal of Philosophy of Education, 37*(3), 241–252.

Payne, S. L. (2000). Challenges for research ethics and moral knowledge construction in the applied social sciences. *Journal of Business Ethics, 26*(4), 307–318. https://doi.org/10.1023/A:1006173106143.

Reckwitz, A. (2002a). The status of the "material" in theories of culture: From "social structure" to "artefacts". *Journal for the Theory of Social Behaviour, 32*(2), 195–217.

Reckwitz, A. (2002b). Toward a theory of social practices. A development in culturalist theorizing. *European Journal of Social Theory, 5*(2), 243–263. https://doi.org/10.1177/13684310222225432.

Reckwitz, A. (2012). Affective spaces: A praxeological outlook. *Rethinking History, 16*(2), 241–258.

Roberts, D. A. (1988). What counts as science education? In P. Fensham (Ed.), *Developments and dilemmas in science education*. London: Falmer Press.

Roberts, D. A. (2011). Competing visions of scientific literacy: The influence of a science curriculum policy image. In C. Linder, L. Östman, D. A. Roberts, P.-O. Wickman, G. Erickson, & A. MacKinnon (Eds.), *Exploring the landscape of scientific literacy*. Oxon: Routledge.

Roberts, D. A., & Bybee, R. W. (2014). Scientific literacy, science literacy, and science education. In N. G. Lederman & S. K. Abell (Eds.), *Handbook of research on science education* (pp. 545–558). Oxon: Routledge.

Sayer, A. (2013). Power, sustainability and well being. An outsider's view. In E. Shove & N. Spurling (Eds.), *Sustainable prctices. Social theory and climate change* (pp. 167–180). Milton Park: Routledge.

Schatzki, T. (1996). *Social practices. A Wittgensteinian approach to human activity and the social*. Cambridge: Cambridge University Press.

Schatzki, T. (2001). Introduction. Practice theory. In T. Schatzki, K. K. Cetina, & E. von Savigny (Eds.), *The practice turn in contemporary theory* (pp. 1–14). London: Routledge.

Schatzki, T. (2013). The edge of change. On the emergence, persistence, and dissolution of practices. In E. Shove & N. Spurling (Eds.), *Sustainable practices. Social theory and social change* (pp. 31–46). Milton Park: Routledge.

Turner, S. (2001). Throwing out the tacit rule book. Learning and practices. In T. Schatzki, K. K. Cetina, & E. Von Savigny (Eds.), *The practice turn in conteporary theory* (pp. 120–130). London: Routledge.

Watson, C. (2011). Staking a small claim for fictional narratives in social and educational research. *Qualitative Research, 11*(4), 395–408. https://doi.org/10.1177/1468794111404317.

Windschitl, M. (2008). Our challenges in disrupting popular folk theories of 'doing science'. In R. A. Duschl & R. E. Grandy (Eds.), *Teaching scientific inquiry*. Rotterdam: Sense.

Gerd Johansen is associate professor in science education at the Norwegian University of Life Sciences where she works in science teacher education. Her main research interest is how knowledge is produced, shared and verified in science education practices. She also works with the intersection of education for sustainable development and technology.

Trine Anker is professor in Religious Studies at MF Norwegian School of Theology, Religion and Society. She is the Head of the Program of Education at the same institution, and involved in teaching and supervising teacher students. Anker has written a Ph.D. about respect among pupils in public schools, and has published articles in national and international journals and anthologies about Religion Education, Professional Ethics and the handling of the terror attack on 22 of July in Norwegian Schools.

Chapter 5
Ethical Considerations in Ethnographies of Science Education: Toward Humanizing Science Education Research

Minjung Ryu

5.1 Introduction

An increasing number of science education researchers employ ethnography as their methodological approach. Ethnographic studies in science education vary widely in their goals, forms, and settings. Some studies aim to understand and document learners and particular learning settings without any intervention while others implement a certain program or curriculum and collect ethnographic data to evaluate their program design, revise it for future implementation, and draw theories about learning. For instance, Heidi Carlone in one study ethnographically examined culture in the classroom of exemplary teachers to understand what afford science identity construction of ethnically and socioeconomically diverse student groups (Carlone et al. 2011) while in another study she and her colleagues implemented a summer science enrichment program and studied its participants' identity work through ethnographic means (Carlone et al. 2015). Some studies collect data through series of interviews and observation (e.g., Long 2012), while others heavily rely on video recording techniques (e.g., Swanson et al. 2014). Furthermore, many scholars conduct ethnographies in community-based organizations beyond formal school and classroom settings (e.g., Carlone et al. 2015; Moje et al. 2004; Rahm et al. 2014). Some studies even actively involve research participants as partners, rather than recipients, and co-develop their research project (e.g., Barton et al. 2016; Irizarry and Brown 2014).

Although research goals and specific data collection and analysis methods differ, there are some commonalities among ethnographic studies in science education. Ethnography in education employs several key data collection methods for an extended period of time, such as "participant observation and and/or permanent

M. Ryu (✉)
University of Illinois at Chicago, Chicago, IL, USA
e-mail: mjryu@uic.edu

© Springer Nature Switzerland AG 2020   71
K. Otrel-Cass et al. (eds.), *Examining Ethics in Contemporary Science Education Research*, Cultural Studies of Science Education 20,
https://doi.org/10.1007/978-3-030-50921-7_5

recordings of everyday life of learners in natural education settings" (Delamont and Atkinson 1995, cited by Gordon et al. 2001, p. 188), as well as semi-structured interviews and casual conversation (Wolcott 1999). Moving away from post-positivistic perspectives, many ethnographers in science education perceive science learning as a sociocultural process that is situated in multi-layered macroscopic and microscopic contexts, and they have a strong commitment to transforming the practices of science education, particularly for underserved learners (Brandt and Carlone 2012). To a varying degree, researchers interact and build relationships with learners, teachers, and/or community members during, and beyond, their research period. These theoretical commitment and methods pose several ethical questions regarding data collection and analysis, relationship-building, and representation of findings.

In this chapter, based on experiences from my research studies (an ethnography in high school biology classes and an after-school science project engaging resettled refugee teens), I discuss ethical quandaries involving three aspects: benefiting study participants and their communities, building relationships with study participants, and constructing stories to tell from the study. I note that these aspects are not simply about research methods, but require ontological, epistemological, and methodological reflections (Murphy and Dingwall 2001). Then, drawing on the ideas of humanizing research (Paris and Winn 2014), I discuss my practices in research to address these ethical concerns. Humanizing research approaches foreground dialogic consciousness-raising for participants and researchers and the building of relationships of dignity, care, and respect between them. Although this methodological stance is important in any research, it is particularly important in research that involves learners oppressed and marginalized by systems of inequality (Paris 2011). Based on the discussion about my own research practice, I will call for researchers' enhanced reflexivity, relevance, and reciprocity throughout a research process especially when working with individuals and communities marginalized on the basis of race, class, and language among others.

5.2 Contexts of My Research

My research interests center on science learning and participation of students of Asian descent in the contexts of the United States (hereafter, *Asian American* students). As Asia is a large continent and represents diverse histories, languages, and economic and political contexts, Asian American students encompass diverse cultural, linguistic, and socioeconomic groups. Some were born in the United States as children of immigrants while others were born in their home countries and moved to the U.S. at varying ages and, thus, are English learners. Some are children of wealthy and highly-educated parents while others live under the poverty line, their parents working in shifts for minimum wage.

In addition, a number of refugees flee from life-threatening living conditions in Asian countries. For instance, a large number of Vietnamese and Cambodians left their home countries in 1970 and Burmese and Bhutanese did so in 2000 (Igielnik and Krogstad 2017). Many of these refugees also include school-aged children. Despite this internal diversity, Asians American students are over-associated with *model minority* and the diversity within the group has often been ignored (Lee 1996). What I am interested in, or concerned about, is this diverse group of students, especially those who are marginalized from school science practices for the reasons of race, language, immigrant status, and socioeconomic backgrounds. Through my research, I would like to empower these youth in pursuing their desires with respect to science learning, such as at the moment of learning in the science classroom or their career paths.

As a researcher living and working in the United States, I observe that Asian American students tend to have limited access to opportunities to say what they think and want to pursue, and they tend not to claim their rights to do so. Often, I notice that dominant groups attribute these observed behaviors of Asian students to their culture, by stating for instance that Asian culture emphasizes respect of teachers and others, or language, by assuming that English proficiency is the only reason for Asians' relative reticence (Ryu 2013a). However, research has shown that Asian American students experience ostracism, bullying, and stereotyping and that these experiences discourage them from talking and interacting with non-Asian students and native English speakers (Chou and Feagin 2008). Education research communities in the United States have overstressed the idea of cultural differences and limited English proficiency (Gee 2001) and tended to explain Asian American students' difficulties in U.S. classrooms as results of cultural differences between Western and Eastern and students' limited English proficiency. As a consequence, the education researchers have limited knowledge about their lived experiences, resources for success, challenges that are attributable to their learning contexts *in the United States*, and strategies in pursuing success in the face of challenges. Against this backdrop, I have strived to make Asian youths' diverse voices heard by teachers and scholars and to encourage Asian American students to view themselves as valuable members of communities and science sense-makers. I have hoped that this would empower Asian American students to say what they think and pursue their learning needs and interests.

The first ethnographic study I conducted was my dissertation, a year-long study that examined Korean transnational high school students' learning, participation, and identity construction in two U.S.-based biology classes. I did not have an intervention plan but intended to understand what was going on in high school science classes, which may or may not encourage transnational learners' participation. During that period, I visited a school 2–3 days per week (approximately 60 visits total), observed and video- and audio-recorded the two biology classes, produced extensive field notes, and interviewed various members of the school (e.g., teachers, students, Korean parents). I even rented a room for 3 months to spend more time in school, interact with study participants more frequently, and get to know the local community environment better. I analyzed the collected data through ethnographic analytic approaches, such as drawing out themes, writing up stories of the participants, and highlighting some participants' stories (Atkinson et al. 2001), as well as through video analytic approaches, such as watching unedited video clips to choose episodes for close discourse analysis and analyzing those selected episodes (Derry et al. 2010). Through this analysis, I showed that Korean transnational students tended to be more reticent than U.S.-born and native English speaking students and that their reticence was partly attributable to their positionings as immigrants, racial minorities and *Others* (Kumashiro 1999), and students of limited English proficiency. I also showed how the youth negotiated their identities to be more conducive to belongingness and academic performance and how such identity negotiation was also intertwined with their gender identity. These positionings and limited participation further limit their access to resources for learning (Ryu 2015a, b).

More recently, I designed and implemented an afterschool program for Burmese refugee youth (mostly high school juniors). The youth in this project had lived in the United States for varying lengths of time, from as short as a few months up to 8 years. Before resettlement in the United States, they were born and had lived in Chin State of Burma/Myanmar and a first asylum country (mostly Malaysia). As a result, the youth speak and read English with varying degrees of proficiency, as well as multiple tribal languages that belong to Chin. In this afterschool program, I aimed to apply research-supported effective teaching approaches—employing funds of knowledge (Moll et al. 1992), responsive teaching practices (Hammer et al. 2012), and bi/multi-lingualism as assets for learning (García 2009)—and document the learning of the participants. I hoped to provide them with opportunities to reason about scientific phenomena and verbally share their ideas, and, in turn, empower them as scientific sense-makers. In doing so, I developed a strong partnership with the local Burmese community center, in which my research team received help and gained insights for working with Chin youth and provided support for the needs of the community center and youth.

5.3 Focus of the Ethical Quandaries

In the process of designing research, collecting and analyzing data, and writing, I have faced several ongoing struggles, conflicts, and feelings of unease due to ethical quandaries. These issues are closely related to the methodology (not methods per se) and epistemology that I brought in to conduct these studies. The following is an excerpt from my recent field notes, which vividly demonstrates these quandaries:

> It was the sense of guilt that pressed me so much while and after completing my dissertation. My participants told me their stories that they do not usually share with others. Some of them showed the weakest, vulnerable part of themselves to me, who had nothing to do with their life. I wrote a dissertation, published several papers, and got a job. I got out a lot from it. But I haven't given anything back to them. It was this sense of guilt. Lately, I have been even more struggling and frustrated that I always end up writing depressing stories about my participants and the world around them—how the society/system is constraining them and how visible and invisible oppression is on them (and us). I didn't want to write such sad stories, yet I couldn't resist seeing them in their stories and in my data. (05/10/2016)

The excerpt shows my sense of guilt that I felt for not giving back to the community as much as I gain from the research (*benefiting study participants and their communities*), relationships I established with the participants (*building relationships with study participants*), and struggles in deciding what to write (*constructing stories to tell from the study*). I felt I was not trained to deal with these ethical tensions nor sure where and how to talk about the feeling of guilt and relationships that may go beyond the traditional researcher-researched relationship. In this section, I explore several concrete examples to elaborate on these three ethical quandaries.

5.3.1 Benefiting Study Participants and Their Communities

Perhaps the most salient issue was unequal benefits distributed between me and the study participants. This tension was bigger in my dissertation. The participants, both teachers and students, let me be in the space of their daily classroom lives, see and even video record them, hear their stories that were sometimes meant for me and in other times were not necessarily meant for me, and write about them. The study was successful from my perspective: I made presentations in conferences and published papers. Yet, I constantly had to ask what benefits the participants and community gained from their participation, if any. Did the study introduce any positive changes to the participants' learning and lives and the school setting? In the IRB application, I wrote that teachers "may have an opportunity to reflect upon [their] teaching experience with regard to education of students with diverse backgrounds and think about ways that [they] serve such students and how [they] may do so in the future," and youth participants "may receive psychological benefit from discussing with a researcher [their] experience of studying science and challenges that [they] are facing in learning science." But how do I know whether the participants indeed benefited from their perspectives, not just from my, somewhat narcissistic, perspective? Do I need some kinds of measurements? Or is it even possible to "measure" it? More fundamentally, what does it even mean to benefit the researched in an ethnographic study? Should it be something tangible and have exchange values? Or can something like becoming a trusting friend "who understand [sic] fully" or "knows what [they] go through" be one? (Paris 2011, p. 139).

In my second study, an intervention study at a community-based afterschool setting that employs ethnographic data collection approaches and techniques (e.g., field notes taking, video-recordings of session meetings, ethnographic interviews), similar questions regarding benefiting the participants and communities arose but in different ways. The afterschool STEM program was in an afterschool enrichment program offered by a Burmese community center. In the enrichment program, the community center staff provided a space for the program, occasional enrichment events, support for college applications, and rides for many youth learners, and coordinated volunteers who offered tutoring to the youth. Throughout the project period, I was given one day per week out of its three-day-per-week program to implement my STEM enrichment program. To implement my STEM curriculum, my research team—myself, faculty co-PIs, two graduate students, and one under-graduate student—worked to run the program throughout an academic year. My research team held an hour and a half long weekly meeting to discuss curriculum implementation, spent a couple more hours revising existing curriculum to respond to progress that the youth made (cf. Hammer et al. 2012), and spent another 2 h actually running the program. This intensive programming made me question what the community center would learn from the study and do after the end of my research study. The center was understaffed for running the STEM program on their own. While I hoped the center could continue the STEM program, or adapt it for their needs, without the work of my research team, providing the STEM program appeared to exceed what they could handle. Thus, the center stopped providing the STEM program after the completion of my project.

In addition, there were often tensions between my research team and community partners regarding what should be done, how it should be done, and what to priori-tize. The Burmese community center was committed to building a self-sufficient and integrated Burmese community through youth education, job placement, and advocacy. One important goal of the youth education program was to increase col-lege admission and graduation rates among Burmese youth. The afterschool pro-gram directly supported this goal by providing various resources for college and scholarship application, tutoring, and other academic help. My STEM program aimed to promote youths' critical STEM literacy (cf. Tan et al. 2012), highlighting literacy skills to analyze STEM knowledge and information and utilize that knowl-edge for making positive changes in their own lives and communities.

While the Burmese community center and I agreed on our goals at the macro level—providing better educational opportunities for youth to be successful indi-viduals in society—there were differences and tensions between us. For instance, my project was not primarily aimed at supporting college admission, but rather engaging the youth in more meaningful STEM learning experiences that applies knowledge for their own purposes. These goals may go beyond or take a slightly different direction than facilitating students at being successful in a school system. Thus, when resources were limited (e.g., support staff, space), needs of the com-munity center took more precedence. Sometimes, our visions were at odds — ten-sions between helping youth complete homework that consisted of drill and busy work versus engaging critical learning practices which may not directly translate to their immediate school performance — and we had to decide what was more impor-tant for the youth in each moment of such tensions. As another example, the Burmese community center encouraged youth to use English *only* as an attempt to facilitate English language learning. This pedagogical approach was at odds with my own that encourages emergent multilingual learners to utilize multilingual resources and translanguage for STEM learning and participation (García 2009).

I do not believe that my research project would necessarily provide more benefits than any other daily practice the participants conduct or should take precedence over their interests and daily practices. Because I am not an insider of the commu-nity and do not participate in their practices, I may not have as much insight as the community members who have developed their insight from first-hand experiences.

Thus, I may not know what are the most important values and urgent needs for the community. Or as many teachers point out about education researchers, I may be more 'idealistic' than practitioners in the sense that I, like many other researchers, may fail to appreciate the complexity of everyday mundane practices to maintain the school and their program and accomplish what they believe to be good for their youth.

How can I negotiate different purposes and perspectives between myself and practitioners and build an authentic and trusting partnership? If they would like to continue the education program that I have designed and implemented, how can it be sustained after the project period without funding and manpower from the university? What should I expect and what not? What can practitioners take away from this work if they do not continue the program we offered? If a research project is not sustainable without support from university-based research and cannot be implemented in conjunction with practitioners' daily practices, what would be the goals of the research? In this case of limited sustainability, is the research ethical?

5.3.2   Building Relationships with Study Participants

In ethnography, building relationships with participants is critical to collecting data and conducting a study. The kinds of data collected depend on the kinds of relationships and trust that researchers establish with participants (Atkinson et al. 2001). Regarding the relationships with her study participants, Ellis (2007) described that members of her research setting "seemed to forget [she] was doing 'research'" and viewed her primarily as "Carolyn, a friend, coming to visit" (p. 6–7). In my research projects, I and my research team built close relationships with study participants over the period of research projects. To a certain extent, my participants seemed to think of me as a friend who understands them (Moje et al. 2004; Paris 2011) or someone that they can trust. My study participants often shared private matters with me, such as their life stories, immigration status (some participants were in the United States without a proper document), migration experiences in which their lives were at stake, struggles in their families, and tensions with friends. Although I had permission, through the institutionalized consent process, to hear their stories and write about them, I questioned whether it was ethical to hear their personal matters to accomplish my research goals. I was not sure if youth participants wanted to share those stories privately with me or disseminate them to a larger audience through my academic writing and presentation. In some situations, as Ellis (2007) wrote, they might have wanted to tell me their stories to share just with me. Indeed, in my dissertation work, while having informal conversations with me, a few students lightheartedly asked me if the conversation would be part of my data and be analyzed. I answered "yes," and some of them walked away.

Relationship building with my study participants posed some ethical questions. Sometimes, I wanted to be a good friend to them. At other times, I had to put back on my researcher hat in the sense that I tried to be close to them to collect more and better data. There were moments that I felt conflicted in thinking what I would do if I were their friend as opposed to a researcher. For instance, when youth participants shared stories of discrimination, bullying, and other conflicts with teachers and their friends, I could not decide whether I should respond from a perspective of friend—as a friend, I may try to be empathetic to them—or adult observer and researcher—as an observer and researcher, I may try to be more neutral and analytical rather than making premature judgments to show empathy to them (e.g., Merriam 1988).

Another point of tension was regarding how much I should extend the relationship with participants outside of the research setting and beyond the duration of a research project. For instance, youth participants often send me Facebook friend requests. Do I accept the requests? If I do, I may invade their privacy (Murphy and Dingwall 2001) or even further breach the confidentiality requirement from the perspective of the Institutional Review Board (IRB), since there is a possibility that my Facebook friends, mostly academics in the field of education, recognize they are my study participants. If I do not, it may hurt their feeling and jeopardize my rapport with them. Sometimes, they tag me in photos that we have taken together. Furthermore, I wanted to meet with them (or they want to meet with me) after the study is completed, just to catch up. This may be a violation of ethics from the IRB's standpoint because a closure of study means no more interaction with study participants. What should I do in these cases? Certainly, my relationships with the participants do not end when my study is completed; the relationships last and can even grow further (Paris 2011). In ethnography, the boundary between research and personal life is not clear, and institutionalized research regulations do not reflect this nature of relationships with study participants. With widespread uses of social media, the boundary has become even more blurry. How do I negotiate this boundary between the researcher-researched relationships and personal relationships in the ways in which no one is harmed and institutionalized regulations are not violated?

The final point of tension is attributable to different kinds of relationships I ended up developing with different groups of participants. In a cultural setting that an ethnographer studies, there may be more than one social group based on different social roles and identities that are ranked differently in a social hierarchy. Their interests are sometimes in agreement and at other times in conflict. Thus, it is inevitable that a researcher would develop different kinds or degrees of relationships with participants because of her own roles and identities. For instance, in schools there are teachers and students from dominant groups (e.g., White, male) and from "minoritized" groups (Harper 2012). Teachers have power and authority over students and also are gatekeepers for my research since they decide whether or not to let me in their classrooms. Let's assume that I want to hear counter-narratives from students that teachers often do not hear or even ignore. What is the relationship that I should aim to build with teachers to get access to the research site while still ensuring that students trust me and share their stories with me? How much does the relationship between the teacher and her students as well as between the teacher and myself impact my relationship with the students?

Even among students, often several different social groups are formed, which assume different degrees of social power and status. In my dissertation study setting (Ryu 2012), several distinct social identities existed based on racial and linguistic identification. Among those groups, I mostly interacted with multi-ethnic old-timer Asians, who immigrated to the United States before elementary school or were children of immigrants, or newcomer Koreans, who moved to the United States relatively recently. While these two groups shared racial and linguistic backgrounds to a certain extent, they typically did not interact across this boundary and occasionally were even in conflict (Ryu 2013b). Although I wanted to be close to both groups as much as I could, it was not always straightforward since my participants identified me as either one or the other. Students seemed to identify me more as a newcomer Korean. Sometimes, in order to be closer to and hear stories of old-timer Asians, I had to make efforts to relate with them. Would it be possible for an individual researcher to interact with multiple social groups equally when she (always) brings her own racial and ethnic (and other complex dimensions of) identity to the research setting? Would it be possible especially in school settings in which students tend to separate themselves from other racial groups and do not socialize across racial borders (Tatum 1997)? Does it mean that the validity of my study is challenged? How do I negotiate and position myself in a research setting? How do I take into consideration my own positionality in the research setting when I analyze the collected data and write?

5.3.3 Constructing Stories to Tell from the Study

As a critically-oriented ethnographer, I tend to see oppressions, constraints, and challenges that systems impose on their members (for a similar example, see Madison 2011). Most of my research participants were positioned as English learners, immigrants, and racial minorities in the U.S. school contexts. To a varying degree, they were marginalized in school classrooms, did not have sufficient opportunities to participate in learning practices as valued and appreciated class members, and were not empowered to seek out and advocate for their learning opportunities. For instance, a Burmese youth who participated in my afterschool program said he spoke English more often in our program than he did in school. It was surprising to me given the fact that all participants in my program were Chin/Burmese and communicated with each other in several common Chin languages, Burmese, and English, while in school he must be surrounded by many monolingual English speakers who communicate in English. He explained that in school he socialized with peers mostly from the same ethnic group as his own and did not talk much to teachers or English-speaking peers and thus he did not have opportunities to use English. His candid remark showed me how he, and probably other Chin/Burmese youth, was positioned and marginalized as an emergent bilingual in school.

Researchers have shown that oppressed and marginalized members of a system try to push the boundaries, negotiate what is possible, and produce counter-narratives (e.g., Marlowe 2010 and Tabar 2007 on the work with refugees, Calabrese Barton and Tan 2018 on the work with youth of color in urban afterschool maker programs). Yet, the boundaries that they push against are situated in a long history of oppression and rigid power structure. Although it is not impossible, the boundaries are not easily altered. Because of the rigid power structure, I typically end up writing depressing stories about my research setting and participants. From my dissertation study, I wrote and published two papers that show how Korean transnational students' status as non-native English speakers, recent immigrants, and Asians positioned them at a lower position in the social hierarchy among students and that this position impeded their opportunities to participate in learning practices and a construction of identity as valuable, capable, and contributing class members. These positionings were further intertwined with their socioeconomic status and gender, wherein youth from lower socioeconomic backgrounds and girls faced more challenges in re-negotiating their positionings. Although my writings were based on the findings of rigorous research and I believe they are trustworthy (Lincoln and Guba 1985), the papers seemed to be telling stories of oppression without presenting a way out.

I question how I can write more promising stories, such as stories of hope and potentials of change. More fundamentally, can and should I write promising stories? Can I refuse to write depressing stories or to disclose potentially depressing findings from my study findings? Is it justifiable or would it merely serve my narcissistic purposes of doing research? Or is there a right balance between revealing injustice and oppression of a system and showing promise and possibilities for its members? How can I change my perspective to write promising stories rather than depressing ones? What does "being ethical" mean in this regard?

5.4 Moving Toward Humanizing Science Education Research

In this section, I intend to present my tentative responses to the ethical quandaries I raised in the previous section drawing on the idea of humanizing research (Paris and Winn 2014). Rather than assuming that these responses are the solutions to my questions, I present them as my current practices and orientations to conducting ethical and humanizing science education research. The humanizing approach takes a methodological stance and process that involves "the building of relationships of care and dignity and dialogic consciousness raising for both researchers and participants" and "reciprocity and respect" (Paris and Winn 2014, p. xvi). Taking the humanizing approach, researchers constantly and carefully reflect on their research practice vis-à-vis their positionality, engage themselves and participants in the collaborative work of liberating participants rather than further marginalizing them, and flexibly adapt to needs and emergent changes in the research context throughout the entire research process rather than merely adhering to a strict research protocol

(Green 2014). This approach has emerged among researchers who work with those who are oppressed and marginalized in the society and schools and encourage researchers and participants to work against inequities "not only through the findings of research but also through the research act itself" (Paris 2011, p. 140).

Humanizing research approach provides valuable insights for pursuing my two research studies presented here because these two fundamentally address the issues of equity and justice in education and aim to empower marginalized learners or learners of marginalized groups. Furthermore, methodological approaches that I employ, that is variations of ethnography, allow flexibility in the research process to realize important ideas of humanizing research. Below I organize my responses to the ethical quandaries by phases of conducting research: research designing, positioning while in the research setting, writing research findings, and leaving the research setting. Although the actual research process is not as linear as it sounds in the phrase *research phases*, I hope this organization may provide a holistic view into what decisions I made throughout the entire research process.

5.4.1 Designing of a Research Project

In designing a research study and setting up goals for the research, I foreground benefits of participants, which include learners, teachers, and community practitioners. This requires a dialogic communication between researchers and potential participants from the designing phase of a research project. In many research projects, researchers set up research goals, design studies, and recruit participants to the fully-designed research project. Instead, from the beginning, researchers and participants should negotiate their visions, goals, and intended actions, since participants should be an important stakeholder in the project. Research goals should go beyond testing researchers' hypotheses, generating new knowledge for themselves, and contributing to the research community. They should consider the praxis— short-term and long-term impact of the research on the setting in which the study is conducted and sustainability of the practice after the research project is completed.

For a professional development project that I am designing currently, I apply these approaches. The project aims to design and offer a professional development program for science teachers who teach in linguistically diverse classrooms. Early in the research design phase, I had meetings with several teachers who were interested in the work. We discussed our visions, goals, and needs of each other. Based on the meetings, I designed a draft of my research design, shared it with them, received feedback, and revised my design. To better understand the resources, needs, and other contexts of the school setting, I plan to conduct an ethnographic study of the school. I expect that this ethnography will help me gain insights into the rich resources that stakeholders can bring, different needs of different stakeholders, and potential benefits that the participants may gain from the research. My goal is to incorporate their expertise and needs and negotiate my goals, as well, within the extent in which the negotiated goals adequately align with the visions of my research.

In addition, I do not assume that a researcher can "save" participants from certain conditions, which implies savior-savage relationship between the researcher and participants (Souto-Manning 2014). Rather, my goal is that both participants and I mutually pursue our own, but negotiated, goals and actively find opportunities for benefiting ourselves and each other. These benefits also are not always defined a priori but rather emerge throughout the research process. In my research projects, I recognized various types of benefits that my study participants have taken from the research and our relationships. Some youth participants in my dissertation seemed to be happy that they could share their stories, that they do not share with other people, with me, who understood their positions and struggles, wanted to hear their stories, and was willing to be their ally. Some participants in my afterschool project asked me for a recommendation letter for their college application. The director of the Burmese community center asked me to serve as a collaborator or advisor for grant proposals the center submitted during and after my research period. The staff said that the partnership with me, a faculty member at a flagship university, provided the center with leverage in increasing visibility of the organization on and off-line (local communities), and that the partnership is one of the key benefits they gained. These benefits were not planned at the project designing phase, but emerged as we, my research team and the participants, moved forward and established a trusting and long-term relationship.

5.4.2   Positioning of the Researcher in the Research Setting

Throughout the process of conducting a research project, I constantly think about my own positionality with respect to the research setting and participants. One important lesson that I learned is an insight into the insider-outsider distinction. This tension between insider versus outsider in a research project has long been discussed in texts in qualitative research methods (Mercer 2007; Naples 1996). While being an insider may help a researcher establish rapport and provide deeper insights into a study setting, it may result in being identified with an identity model (Wortham 2006) that is employed to characterize certain members in the study setting. As discussed earlier, in my dissertation study setting I was mostly identified as a newcomer Korean. This, in turn, generated unwanted tensions and conflicts with participants who identified with a different identity model.

Many scholars have challenged this insider and outsider dichotomy and that one status is better than the other (e.g., Mercer 2007). Rather, a researcher's position with respect to study participants changes dynamically based on multiple social dimensions and offers different kinds of affordances. Beyond the insider-outsider distinction, Kinloch and San Pedro (2014) redefined the ethnographic research process as actively collaborating with participants to learn about complexities in which the lives of participants are situated. In this conceptualization of research, research is not simply reporting insiders' stories, and the distinction between insider and outsider is less meaningful. Individuals—participants and researchers—are unique and agentively make sense of each other. The research is non-linear and a reciprocal process of engaged listening and dialogic sense-making of each other (Forsey 2010). Telling stories and taking actions for meaningful changes, based on this sense-making, are key to ethical and humanizing research practices.

I believe that instead of trying to achieve the insider status, a researcher should critically reflect on her own different and shifting positions with respect to individual participants in the research setting and maintain somewhere in between an outsider and insider status, perhaps a close outsider who has more understanding about the setting than other far outsiders and provides a connection between the research setting and those far outsiders—perhaps "allied others" (McCarty et al. 2014). One's positionality does not simply mean what a researcher thinks she is, but includes how research participants view who the researcher is in relation to themselves. Recognizing her own positions encourages a researcher to be more conscious about her perspectives and approaches throughout interactions with the participants and data collection and analysis. Further, a researcher should be attentive to her positions impacted by ongoing social and political dynamics among community members and capitalize on the different positions.

Thus, while I tried to interact with participants and be close to them, I also maintained a certain distance from them in order to present myself flexibly in different interactional situations. Although identity models circulating in the research setting affected my positions, I tried not to be cast into one of them. By being an outsider to a certain extent, I could make conscious decisions as to how to navigate my identity space and negotiate my multiple identities depending on interactional contexts. In addition, when a research team is composed of multiple members, their different relationships can also be utilized. For instance, in the afterschool project, my two graduate students and I had various levels of interaction with different youth participants, and, in turn, established different relationships. Instead of being equal to every youth participant, we purposefully chose who would interview each student based on our relationships and rapport with students.

In analyzing collected data and writing my analysis, I try to be cognizant of my own positioninings that are reciprocally constructed in the research setting and make my presence transparent. In writing about data collection, I treat myself as a live participant and agentive sense-maker in the research site, not a relation-free and interaction-free data collector. In every interactional moment, both the participants and I make sense of each other, interpret each other's utterances and behaviors, and respond accordingly, which became my data. This process is also reciprocal and dialogic in that one's sense-making and response impact those of the other. When analyzing the data, I often had to infer what the participants' sense-making was about any given interaction. Sometimes, such inferences are made with direct evidence and at other times without direct evidence, which is what social beings do in daily interactions. Instead of staying away from making those inferences, I try to elaborate how and on what ground I make a certain inference as an ethnographer who has studied the setting for an extended period.

5.4.3   Writing as an Ethical Conduct of Research

Reporting research findings through conference presentations and publications requires ethical consideration. First, while research participants agree to participate in the research and voluntarily share their stories with me, they may not always be conscious of the fact that I may write about their stories and publish them. Ellis (2007) wrote about a situation in which her participants later read her writing and became upset because she disclosed some secrets shared between the participants and her. Some data that I collected could be private and potentially sensitive to the participants, such as family crises, relationships with their peers, and immigrant status. Among such sensitive stories, some were useful and worth citing because the data provided detailed contexts of the participants' lives and supported my claims. When encountering such situations, I carefully evaluate whether that piece of data is crucial to understanding the participants and making my claims. I ask myself if I want to use the information to dramatize or romanticize the participants' stories. Then, I cite sensitive information only if I cannot make my claim without the information. If possible, I try to discuss with my participants about disclosure of specific information or find other pieces of data to support my claim that seem to be less sensitive.

Second, I try not to write depressing stories about my participants and stories that potentially lead to stereotyping them. As Simpson (2007) critically asked, I question myself, "Can I do this and still come home; what am I revealing here and why? Where will this get us? Who benefits from this and why?" (p. 78). Tuck and Yang (2014) argued that social science research elicited "pain stories from communities that are not White, not wealthy, and not straight" (p. 227). These collective pain stories often address abuse of power, oppression imposed by the system, and pains of the oppressed. Tuck and Yang argued that while many scholars choose other lines of inquiry, novice researchers tend to write pain stories prematurely believing that it represents what social science research is and does. Instead of pain stories, they suggested desire-based research and pursuing a more complex and dynamic understanding of individuals and communities in real lives. They argued that refusal to write pain stories is "not just 'no,' but a redirection to ideas otherwise unacknowledged and unquestioned" (p. 239). I push myself to ask this question when I am tempted to focus on pains of my participants and turn my perspective around to look at different aspects of their lives and learning.

I admit that my dissertation was focused much on pain stories. It was partly because of the nature of the research questions that I asked ("How are Korean immigrant students' identities enacted in science classroom settings?") and partly because of my intention in the study to challenge the model minority stereotype about Asian Americans (Lee 1996) and report the stories of tensions and struggles that Asian American students experience, which have not been much known in the literature. In my recent studies, I try to focus on the details of individuals' diverse and complex stories, as a way to empower the participants and communities, and look for stories of negotiation, change, and hope, as a way to move away from writing collective pain stories. For instance, in my recent publication (Ryu and Tuvilla 2018), I described how resettled refugee youths were positioned as ethnic, racial, and linguistic minorities and former refugees *and* how they negotiated their identity narratives by providing diverse and empowering stories of themselves and their communities. My goal was to demonstrate empowering stories that the youths authored, by highlighting evidence of hope, desire, and future change. Instead of ending my writings with pain stories, I focused on their agency and resilience as an agent of their own life and learning. I hoped to shed light on constant push and pull between the oppressive structure and agency and, by doing so, disrupt dominant pain (only) stories of the refugee youth and contribute to humanizing research practices.

5.4.4 Leaving the Research Setting

Figueroa (2014) raised an issue regarding how ethnographers leave a research setting; that is, ethics of departure. She pointed out that graduate programs that train qualitative researchers focus heavily on entry into the research setting, but rarely on departure from a research setting and closure of a study: What role should a researcher take with respect to individuals and communities after the closure of a study? How should, or should not, relationships be maintained between the researcher and the researched? She suggested that departure should be more explicitly discussed throughout the ethnographic studies and be included as a key milestone of dissertation writing. Discussions about departure, she argued, should include strategies and politics of maintaining or breaking relations with the participants and their communities.

As Figueroa implied, there may not be one correct way to handle departure from a research setting. Instead, a researcher should reflect on her own goals and positions, those of participants, and reciprocity of the relationships between the researcher and participants. For my two research projects, I try to maintain the relationship with my study participants, including adults, teachers, and youth participants. This means I make myself available for them and I occasionally contact them to say hello. A recent advancement in social media, such as Facebook and applications for mobile devices, makes it easier to maintain such relationships. Immediately after my departure, I stayed in touch with some participants, who I developed a closer rapport than I did with others, and met a couple of participants because we were genuinely curious about how each other was doing. Through the social media connections, I occasionally requested follow-up interviews and/or asked questions for my data analysis. The participants asked me similar favors, such as writing a recommendation letter for college applications.

Through the sustained relationship, I try to share papers that I publish from research projects. I share published papers for several reasons. First, I would like to be as open as possible regarding what I do with data about the participants. Second, I would like them to recognize how much their stories are important and have potential to make changes to education. Third, the fact that my study participants would read my papers encourages me to think more carefully about how to represent their stories. Finally, providing them an opportunity to read my papers may benefit the participants and community. Perhaps they may recognize in my paper what they did not know about their own practices and experiences or appreciate that their voices are heard by others. Typically, it takes a few months to years to publish findings from an ethnographic study. For my dissertation study, the time lag between the data collection and publications was as long as 4 years. Thus, maintaining a relationship helps to share my papers after some time has passed since the closure of the study.

For instance, 3 years after I left my dissertation research setting, I met with the classroom teacher, with whom I became friends on Facebook, talked briefly about the findings of my work, and gave papers that I had published in case she wanted to read them. I admit that it was not easy and somewhat awkward to ask for a meeting to share my papers and tell her about my suggestions based on the study findings, especially a while after the closure of the study. Nonetheless, the teacher was positive and curious about my thoughts and talked to me about her own learning during and after the period of my research. After the meeting, I was somewhat relieved from the sense of guilt that my study did not contribute to the participant community.

5.5 Conclusion

In this chapter, I discussed ethical quandaries that I have faced in my science education research projects that adopted ethnographic approaches, and practices and principles that I employ to handle the ethical issues. There may not be one right way to resolve ethical issues arising in ethnographic research. It is also unrealistic that all ethical issues can be regulated through institutionalized practices, such as the Institutional Review Board. Monitoring every single component of research practice might rather restrict possibilities for challenging the dichotomy between the researcher and researched and over-rate "scientific" and "objective" research in education research. Instead, researchers' reflexivity, relevance, and reciprocity should be more emphasized throughout the process of conducting research (Green 2014). These three principles include researchers' critical reflection on all dimensions of the project, relevance of the research project to the participants and community, and reciprocal relationship and mutual benefits between the researcher and participants. Through such research practices, a researcher may be able to navigate and negotiate various ethical quandaries emerging in various research settings. In this way, science education research communities may move a step forward to humanizing research, away from conducting research dehumanizing humans.

References

Atkinson, P., Coffey, A., Delamont, S., Lofland, J., & Lofland, L. (2001). *Handbook of ethnography*. Thousand Oaks: Sage.

Barton, A. C., & Tan, E. (2018). *STEM-rich maker learning: Designing for equity with youth of color*. New York: Teachers College Press.

Barton, A. C., Tan, E., & Greenberg, D. (2016). The makerspace movement: Sites of possibilities for equitable opportunities to engage underrepresented youth in STEM. *Teachers College Record, 119*(6), 11–44.

Brandt, C. B., & Carlone, H. (2012). Ethnographies of science education: Situated practices of science learning for social/political transformation. *Ethnography and Education, 7*(2), 143–150.

Carlone, H. B., Haun-Frank, J., & Webb, A. (2011). Assessing equity beyond knowledge-and skills-based outcomes: A comparative ethnography of two fourth-grade reform-based science classrooms. *Journal of Research in Science Teaching, 48*(5), 459–485.

Carlone, H. B., Huffling, L. D., Tomasek, T., Hegedus, T. A., Matthews, C. E., Allen, M. H., & Ash, M. C. (2015). 'Unthinkable'Selves: Identity boundary work in a summer field ecology enrichment program for diverse youth. *International Journal of Science Education, 37*(10), 1524–1546.

Chou, R. S., & Feagin, J. R. (2008). *The myth of the model minority: Asian Americans facing racism*. Boulder: Paradigm Publishers.

Delamont, S. & Atkinson, P. (1995). *Fighting familiarity: Essays on education and ethnography*. Cresskill, NJ: Hampton Press.

Derry, S. J., Pea, R. D., Barron, B., Engle, R. A., Erickson, F., Goldman, R., et al. (2010). Conducting video research in the learning sciences: Guidance on selection, analysis, technology, and ethics. *The Journal of the Learning Sciences, 19*(3), 3–53.

Ellis, C. (2007). Telling secrets, revealing lives: Relational ethics in research with intimate others. *Qualitative Inquiry, 13*(1), 3–29.

Figucroa, A. M. (2014). La carta de responsabilidad: The problem of departure. In D. Paris & M. T. Winn (Eds.), *Humanizing research: Decolonizing qualitative inquiry with youth and communities* (pp. 129–146). Thousand Oaks: Sage.

Forsey, M. G. (2010). Ethnography as participant listening. *Ethnography, 11*(4), 558–572.

García, O. (2009). *Bilingual education in the 21st century: A global perspective*. Malden: Wiley-Blackwell.

Gee, J. P. (2001). Identity as an analytic lens for research in education. *Review of Research in Education, 25*, 99–125.

Gordon, T., Holland, J., & Lahelma, E. (2001). Ethnographic research in educational settings. In P. Atkinson, A. Coffey, S. Delamont, J. Lofland, & L. Lofland (Eds.), *Handbook of ethnography* (pp. 188–203). Thousand Oaks: Sage.

Green, K. (2014). Doing double dutch methodology: Playing with the practice of participant observer. In *Humanizing research: Decolonizing qualitative inquiry with youth and communities* (pp. 147–160). Thousand Oaks: Sage.

Hammer, D., Goldberg, F., & Fargason, S. (2012). Responsive teaching and the beginnings of energy in a third grade classroom. *Review of Science, Mathematics and ICT Education, 6*(1), 51–72.

Harper, S. R. (2012). Race without racism: How higher education researchers minimize racist institutional norms. *The Review of Higher Education, 36*(1), 9–29.

Igielnik, R., & Krogstad, J. M. (2017). *Where refugees to the U.S. come from*. Retrieved Feb 27 2019, from Pew Research Center. http://www.pewresearch.org/fact-tank/2017/02/03/where-refugees-to-the-u-s-come-from/

Irizarry, J., & Brown, T. (2014). Humanizing research in dehumanizing spaces: The challenges and opportunities of conducting participatory action research with youth in schools. In D. Paris & M. T. Winn (Eds.), *Humanizing research: Decolonizing qualitative inquiry with youth and communities* (pp. 63–80). Thousand Oaks: Sage.

Kinloch, V., & Pedro, T. S. (2014). The space between listening and storytelling: Foundations for projects in humanization. In D. Paris & M. T. Winn (Eds.), *Humanizing research: Decolonizing qualitative inquiry with youth and communities* (pp. 21–41). Thousand Oaks: Sage.

Kumashiro, K. K. (1999). Supplementing normalcy and otherness: Queer Asian American men reflect on stereotypes, identity, and oppression. *International Journal of Qualitative Studies in Education, 12*(5), 491–508.

Lee, S. J. (1996). *Unraveling the "model minority" stereotype: Listening to Asian American youth.* New York: Teachers College Press.

Lincoln, Y. S., & Guba, E. G. (1985). *Naturalistic inquiry.* Beverly Hills: Sage.

Long, D. E. (2012). Evolution education in policy and practice: An ethnographic perspective. *Ethnography and Education, 7*(2), 197–211.

Madison, D. S. (2011). *Critical ethnography: Method, ethics, and performance.* Thousand Oaks: Sage.

Marlowe, J. M. (2010). Beyond the discourse of trauma: Shifting the focus on Sudanese refugees. *Journal of Refugee Studies, 23*(2), 183–198.

McCarty, T. L., Wyman, L. T., & Nicholas, S. E. (2014). Activist ethnography with indigenous youth: Lessons from humanizing research on langauge and education. In D. Paris & M. T. Winn (Eds.), *Humanizing research: Decolonizing qualitative inquiry with youth and communities* (pp. 81–104). Thousand Oaks: Sage.

Mercer, J. (2007). The challenges of insider research in educational institutions: Wielding a double-edged sword and resolving delicate dilemmas. *Oxford Review of Education, 33*(1), 1–17.

Merriam, S. B. (1988). *Case study research in education: A qualitative approach.* San Francisco: Jossey-Bass.

Moje, E. B., Ciechanowski, K. M., Kramer, K., Ellis, L., Carrillo, R., & Collazo, T. (2004). Working toward third space in content area literacy: An examination of everyday funds of knowledge and discourse. *Reading Research Quarterly, 39*(1), 38–70.

Moll, L. C., Amanti, C., Neff, D., & Gonzalez, N. (1992). Funds of knowledge for teaching: Using a qualitative approach to connect homes and classrooms. *Theory Into Practice, 31*(2), 132–141. https://doi.org/10.1080/00405849209543534.

Murphy, E., & Dingwall, R. (2001). The ethics of ethnography. In P. Atkinson, A. Coffey, S. Delamont, J. Lofland, & L. Lofland (Eds.), *Handbook of ethnography* (pp. 339–351). Thousand Oaks: Sage.

Naples, N. A. (1996). A feminist revisiting of the insider/outsider debate: The "outsider phenomenon" in rural Iowa. *Qualitative Sociology, 19*(1), 83–106.

Paris, D. (2011). 'A friend who understand fully': Notes on humanizing research in a multiethnic youth community. *International Journal of Qualitative Studies in Education, 24*(2), 137–149.

Paris, D., & Winn, M. T. (2014). *Humanizing research: Decolonizing qualitative inquiry with youth and communities.* Thousand Oaks: Sage.

Rahm, J., Lachaîne, A., & Mathura, A. (2014). Youth voice and positive identity-building practices: The case of science girls. *Canadian Journal of Education, 37*(1), 1.

Ryu, M. (2012). Revisiting the silence of Asian immigrant students: The negotiation of Korean immigrant students' identities in science classrooms (Publication No. 3543605). [Doctoral dissertation, University of Maryland, College Park]. ProQuest Dissertations Publishing.

Ryu, M. (2013a). "But at school … I became a bit shy": Korean immigrant adolescents' discursive participation in science classrooms. *Cultural Studies of Science Education, 8*(3), 649–671.

Ryu, M. (2013b). Dancing between Kyung Soo and Mike: Identity formation through dialogic interactions. In R. Endo & X. L. Rong (Eds.), *Educating Asian Americans: Achievement, schooling, and identities* (pp. 181–204). Charlotte: Information Age Publishing.

Ryu, M. (2015a). Positionings of racial, ethnic, and linguistic minority students in high school biology class: Implications for science education in diverse classrooms. *Journal of Research in Science Teaching, 52*(3), 347–370.

Ryu, M. (2015b). Understanding Korean transnational girls in high school science classes: Beyond the model minority stereotype. *Science Education, 99*(2), 350–377.

Ryu, M., & Tuvilla, M. (2018). Resettled refugee youth's stories of migration, schooling, and future: Challenging dominant narratives about refugees. *The Urban Review, 50*(4), 529–558.

Simpson, A. (2007). On ethnographic refusal: indigeneity, 'voice' and colonial citizenship. *Junctures: The Journal for Thematic Dialogue, 9*.

Souto-Manning, M. (2014). Critical for whom? Theoretical and methodological dilemmas in critical approaches to language research. In D. Paris & M. T. Winn (Eds.), *Humanizing research: Decolonizing qualitative inquiry with youth and communities* (pp. 201–222). Thousand Oaks: Sage.

Swanson, L. H., Bianchini, J. A., & Lee, J. S. (2014). Engaging in argument and communicating information: A case study of English language learners and their science teacher in an urban high school. *Journal of Research in Science Teaching, 51*(1), 31–64.

Tabar, L. (2007). Memory, agency, counter-narrative: Testimonies from Jenin refugee camp. *Critical Arts, 21*(1), 6–31.

Tan, E., Barton, A. C., Turner, E., & Gutiérrez, M. V. (2012). *Empowering science and mathematics education in urban schools*. Chicago: University of Chicago Press.

Tatum, B. D. (1997). *Why are all the black kids sitting together in the cafeteria? And other conversations about race*. New York: Basic Books.

Tuck, E., & Yang, K. W. (2014). R-Words: Refusing research. In D. Paris & M. T. Winn (Eds.), *Humanizing research: Decolonizing qualitative inquiry with youth and communities* (pp. 223–247). Thousand Oaks: Sage.

Wolcott, H. F. (1999). *Ethnography: A way of seeing*. Lanham: Rowman Altamira.

Wortham, S. E. F. (2006). *Learning identity: The joint emergence of social identification and academic learning*. New York: Cambridge University Press.

Minjung Ryu is an assistant professor in Chemistry and Learning Sciences at University of Illinois at Chicago, USA. Her research focuses on STEM learning and participation of cultural and linguistic minority students. Employing ethnography and discourse analysis, she examines how racial, ethnic, and linguistic minority students engage in STEM discourses using multilingual and multimodal means and what are ways to design learning environments to improve these students' learning experiences. Within this research interest, she has worked with resettled Burmese refugee teens in a community-based afterschool program in USA where the teens learn STEM knowledge to transform their communities and global societies. She also has collaborated with high school science teachers to develop instructional materials and practices to support English learners in linguistically superdiverse classrooms. Minjung has published in *Journal of Research in Science Teaching, Science Education*, and *International Journal of Science Education*.

Chapter 6
Sex Education—Normativity and Ethical Considerations Through Three Lenses

Auli Arvola Orlander and Iann Lundegård

6.1 Focusing on Ethical Concerns with the Lens as a Metaphor

Viktor: Usually, only vaginal sex is regarded as sex … the norm is that a woman gets her orgasm through vaginal sex …
Ulrika: Yes.
Viktor: … and that's the only thing that gives her pleasure.
Ulrika: Exactly! And the norm is that heterosexual sex is what's normal, so to speak.
Viktor: It is our macho culture that has created it.

All societal discourses, including school and education discourses, are entangled in values and norms. Irrespective of whether the educational content is drama, language or science, some specific value-based dimensions are highlighted as more important than others. Even when the students themselves are asked to discuss an issue, certain normative assumptions are given higher priority. It is important to pay attention to and discuss this tendency in the context of teaching as well as in the educational research that examines such teaching. In the short example above two upper secondary students, Ulrika and Viktor, point out what they identify as common sexuality and relationship assumptions in contemporary society. Together they identify common norms and ethical considerations they regard as being typically included in a discourse about human sexuality in society and in education: that sex implies penetration (and is that which gives a woman an orgasm), that sexuality should be defined on the basis of heterosexual relationships, and that the premises for these norms are set by a male-oriented society. The conversation between the

A. A. Orlander (✉) · I. Lundegård
Stockholm University, Stockholm, Sweden
e-mail: auli.arvola.orlander@mnd.su.se

© Springer Nature Switzerland AG 2020
K. Otrel-Cass et al. (eds.), *Examining Ethics in Contemporary Science Education Research*, Cultural Studies of Science Education 20,
https://doi.org/10.1007/978-3-030-50921-7_6

two students takes place in a classroom where they were explicitly assigned the task of identifying and discussing issues relating to sexuality which are taken for granted in our society.

This chapter highlights how ethical norms concerning human sexuality generally and women's bodies specifically manifest when we explore teaching through different lenses. Later, it also becomes clear how great the responsibility is that we as researchers carry when we want to make these conditions visible, since norms and forms of oppression are often situated in a contemporary context. While in some cases research ethics is limited to considerations only of those who participate in a specific study, we want to emphasise the importance of professional ethics. By this we mean "the researcher's responsibility towards research and the research community /.../ Issues of the researcher's behaviour in various roles, of responsibility in connection with publication, and of so-called research misconduct belong to this category." (Swedish Research Council 2017, p12).

To highlight the ethics in a classroom context we draw on the concept of transaction that the philosopher John Dewey developed during his lifetime, and in his final publication eventually refined together with his colleague, the political philosopher Arthur F. Bentley (Dewey and Bentley 1960). From the moment we are born, they claim, our lives unfold in a flow of actions in a certain environment. Instead of assuming the individual as the given object in analysis of knowledge production, their focus shifts to highlight ongoing encounters and actions within them as the object of analysis. These transactional events can then be analysed within different depths of field (DOF) depending on which encounters need to be focused on. Thus, it is possible to shift the analytical focus from one particular action and encounter to another taking place at the same time and in the same activity but from another depth perspective, where one is foregrounded and the other backgrounded. For example, the focus can be shifted from what is revealed by a student's individual reflection (depth of field one, DOF 1), to what comes up when a group of students take part in a conversation (depth of field two, DOF 2), or to what transpires in these narrow settings viewed in the context of the historical and social conditions in which they take place (depth of field three, DOF 3). The latter analysis may also be derived from what the philosopher Foucault (2002) came to call genealogy, simply described as the contingent movement of values and ethics that determines the boundaries of thought and morality in a certain domain and period — a historical and social context that, from a research ethics perspective, becomes crucial to consider.

In order to highlight this shift, from studying values and ethics from an individual perspective, to studying them as they develop in different transactions, we need a new metaphor. Here we use the terms foreground and background and the metaphor of lens. Rogoff (1995) explains how these lenses always occur as mutually dependent in the formation of an activity.

> Nonetheless, the parts making up a whole activity or event can be considered separately as foreground without losing track of their inherent interdependence in the whole. Their structure can be described without assuming that the structure of each is independent of that of the others. Foregrounding one plane of focus still involves the participation of the backgrounded planes of focus (Rogoff 1995, s.140).

In the science of optics, we use various technical lenses to approach an object or phenomena from different depths of field. A stronger lens zooms in and distinguishes detail that a weaker lens cannot. A wide-angle lens reveals larger entities and relationships that the sharper lens cannot. If we apply this metaphor to the concept of transaction, it is not about lifting individual details out of an event, but rather gaining an understanding of phenomena taking place within different depths of field in the same activity, placing some issues in the foreground, others in the background. In the present context, this metaphor helps us when selecting events as they occur at different depths of transaction in a field that deals with teaching about sexuality and relationships.

6.1.1 Companion Meanings

In science education included and unfounded norms, or 'extra' meanings often resulting from what is not discussed, are sometimes talked about as "companion meanings" (Östman 1998). All teaching, it is thus said, comprises a companion meaning—a hidden message that we are not aware of. For example, when a biology teacher organises plants and animals in a food chain or as a trophic pyramid, it may imply that biology, as science, provides a true picture of how nature is construed. Presenting biology as an objective stance then becomes a value in itself. When another teacher in the same discipline uses biological knowledge to demonstrate how ocean oxygen interacts with the same chains and pyramids, and how this is critical for animal and human survival on earth, then the knowledge renders another kind of value for the students. The latter context includes a companion meaning that influences the students' view on the utility of biology in societal issues, while the former claims an objective, factual description. Consequently, there are normative implications that give rise to ethical considerations about what is to be regarded as good, right or beautiful in all teaching. Often, this value-biased content is based on unreflectad habits, which have rarely, or perhaps never, been taken into consideration (Dewey 1957). Sometimes, however, it is useful to raise these unconsidered habits of teaching to the surface and examine what consequences such way to prioritise renders in teaching as well as in research. It was such a task that preceded the discussion between the students above, and also that to which we pay more detailed and closer attention in this chapter when using depth of field (DOF) as a heuristic. Before we proceed with the analysis, we describe the context in which the empirical data were recorded.

6.2   Five-Week Visit to a Science Education Classroom

In this chapter we use events from a science education classroom to highlight the ethics questions that come up in different depths of field while observing 56 students studying sex education. The students had given written permission to follow their work, all in accordance with current ethical regulations (Swedish Research Council in 2017). In a collaborative project between the biology teacher, Monica, and Auli (one of the authors), two upper secondary classes were followed for 5 weeks. During this time, different types of teaching took place. The overall theme of the students' work, as the teacher labelled it, was an examination of a "critical Review of Sexuality" and their task was to norm-critically examine something concerning human sexuality that they regarded as taken for granted in society. Or, as the teacher expressed it, to "search for norms on sexuality that you perceive as present in your everyday life". Moreover, to discuss what kind of consequences these norms could give rise to and how some norms could be challenged with the help of further knowledge of norm criticism and biology.

The data set consisted of several hours of audio recordings of student discussions, recorded student interviews and written submissions with examples of critical studies of norms. The chosen excerpts which demonstrate the analysis work with the three different depth perspectives illuminate not only the phenomena moving through different fields in an activity, but also the ethics that are evident in the students' discussions.

6.3   Ethics in Three Depths of Field (DOF)

In order to illustrate how norms in teaching can be highlighted by shifting focus we use the metaphor of lens and the associated concept, depth of field. We use three depths to zoom in and out on what takes place in different encounters. The first depth of field, DOF 1, focuses on the transaction taking place when an individual student is given the opportunity to take a step back and challenge common assumptions about sexuality and relationships. Here we are able see what kind of framework the individual student constructs when given an opportunity to make a critical analysis of contemporary norms. Thus, the data involve the student's individual reflections and the statements they make before they begin a discussion with their classmates. The second depth of field, DOF 2, focuses on what happens within the immediate exchange of views in the encounters between the students related to the issues they raised in DOF 1. The data consist of all exchanges of content and values between the students when they were involved in conversation.

Finally, we zoom out. Thus, the third depth of field, DOF 3, focuses on what becomes visible when, as researchers, we highlight the transaction between what happens in the classroom and the unspoken historical and societal context within which this takes place (Foucault 2002) and what then, from a research ethics

Table 6.1 A methodological heuristic describing the ethical concerns in three depths of field

Depth of field	Transactions to be studied and highlighted
DOF 1	Ethical concerns raised in transactions between an individual student and the content.
DOF 2	Ethical concerns raised in the transactions between students involved in a communicative activity.
DOF 3	Ethical concerns raised in the transactions between the communicative activities being studied and the historical and societal context within which these activities take place and the researchers' responsibility to shed light on this.

perspective, becomes important to take into account. In Table 6.1 we present a summary of the ethical concerns in the DOFs.

These three different lenses each bring different issues into focus and we need to be aware that it is always we as researchers who must take responsibility for the analytical tools we shape. The use of other "instruments" might have illuminated other patterns in the student discussions. Or, as Donna Haraway (1988) more poetically expresses it,

> There is no unmediated photograph or passive camera obscura in scientific accounts of bodies and machines; there are only highly specific visual possibilities, each with a wonderfully detailed, active, partial way of organizing worlds. All these pictures of the world should not be allegories of infinite mobility and interchangeability but of elaborate specificity and difference and the loving care people might take to learn how to see faithfully from another's point of view /.../ (p 583).

Below we present three norm-critical considerations that the individual students themselves made in relation to the content (DOF 1). Thereafter we show what happened in the communicative exchanges in accordance with these considerations (DOF 2). Each example is followed by a brief summary of the ethical questions the students touched upon in conversation (Tables 6.2, 6.3 and 6.4). Finally, the third lens, DOF 3, focuses on what becomes visible when highlighting the content of the student encounters in relation to the historical and societal context within which they take place.

6.3.1   Example 1

6.3.1.1   DOF 1: "Men have greater sexual desire than women"

The teacher, Monica, has divided the class into small groups consisting of 4–6 students. In one of the groups they discuss Alicia's norm-critical exposition (DOF 1): "Men have a greater sexual desire than women".

Alicia begins by explaining how she conducted her investigation into how sexual desire is said to work and if there really are any relevant biological differences between men and women. She has looked at the norms that might maintain such an

Table 6.2 Summary of ethical issues. Example 1

Ethical concerns that the students touch upon in the norm-critical investigation of the topic "Men have greater sexual desire than women" and "Men are dangerous"
– Should we assume that there is a biological difference between girls' and boys' sexual pleasure, or is it something you learn?
– Should we regard women's sexual desire as shameful?
– Should we assume that there are hormones like testosterone and oxytocin that affect the sex drive of men and women?
– Should sex drive be regarded as something normal, natural and good for men?
– Is it good for society that men sow their seed?
– Should we think that men are just looking for sex?
– Should men who have no sexual drive turn to pharmaceutical companies?
– Do drug companies benefit from these norms?
– Does the norm relating to men having a strong sex drive normalise rape?
– Should one generalise, or is everything individual?
– Should girls be afraid of guys?
– Should all guys be horny?
– Can one generalise that all boys are dangerous because 99% of all rapes are committed by men?
– Should one have sex even if the girl does not want it?
– Should the erection be seen as an enabler of sex, or can a woman rape a man even if he doesn't have an erection?

Table 6.3 Summary of ethical issues. Example 2

Ethical concerns that the students touch upon in the norm-critical investigation into "The length of the act of penetration is important" and "A woman should come through vaginal sex"
– Should sex only be defined by penetration or should the whole act be included?
– Is there a norm that says that a woman should have an orgasm only through vaginal sex?
– Should only penetrative sex between men and women be counted/are homosexuals and people with several partners then excluded?
– Should orgasm be regarded as the main aim of sex/can touching also be counted as sex?
– Should there be a limit for what is counted as sex?
– Must the man delay orgasm in order to please the woman?
– Must the woman have an orgasm several times for it to be considered good

assertion and from which individuals and groups may derive benefit or disadvantage. First, she points out that it has not been an easy task to find reliable sources.

Alicia: I have not found very good sources, but I have found a few on the Karolinska Institute [a medical university] website. There was a professor in clinical sexology who claimed that no actual difference is supported by research. Anyway, boys and girls learn from early ages that the sexual desire of a woman is shameful. Then a woman can … yes you simply get such an idea.

Table 6.4 Summary of ethical issues. Example 3

Ethical concerns that the students touch upon in the norm-critical investigation into "It is going to hurt and will bleed" and "It's important to show that you are a virgin"
– Should girls bleed, and should it hurt during first intercourse?
– Should the maidenhead burst during first intercourse?
– Should we use the word 'maidenhead' or 'vaginal corona'?
– Does the claim that women should bleed during first intercourse cause problems?
– Should some cultures and religions be allowed to regard it as important that girls bleed during first intercourse?
– Should girls be worried that it will bleed and hurt?
– Should doctors in Sweden perform surgery on girls to enable them to bleed during first intercourse?
– Should girls have to bleed to show their virginity?
– Is it important to keep the myth alive?
– Can an operation be a viable way to help girls?
– Should doctors advise girls on how to make it appear as if the hymen had burst, or is it better for doctors to provide men with information about how it actually works?
– Is there a risk that the myth will persist if one fails to provide facts?
– Should Swedish hospitals consider how things are done in other parts of the world?
– Should well-known clinics practice such surgery, or is it just shady clinics that should perform them?
– Should women decide whether they want surgery, or should others decide for them?
– Should it hurt for the woman during first intercourse?
– Do girls need to be limited by the threat that it's going to hurt?
– Can you see if a girl is a virgin?
– Should a girl be a virgin when she has intercourse for the first time?
– Should a girl prove that she is a virgin by showing that she has got rid of something that was never there?

Anna: Yes, it feels like it's a norm somehow.

Alicia: Then how it works, the sexual drive … Here I found some … it was … in the brain … there are a lot of neurotransmitters operating. Among other things, a substance that regulates serotonin, and hormones such as testosterone and one named oxytocin affect sexual desire in women. And what are the norms behind …? It's regarded as good or natural for men, though … because they are going to spread their semen, and that's good for society. Other norms are also in circulation, like, men are just searching for sex. Those who benefit from these norms are … What I most thought about was drug companies. Men who feel that they don't have this sexual desire should be looking for medication, because it's not considered normal to not have a strong sex drive. This explanation has also become a justification for men's sexual behaviour. I also thought about that when it comes to a situation of rape there is so much focus on women's behaviour. Perhaps because it is somehow normalised that men would have this awesome sex drive.

Here, Alicia refers to several norms, which, based on her individual reflection, relate to differences between men's and women's sexual activity — how women's sexuality can be seen as shameful, and how men are naturally expected to have a stronger libido.

6.3.1.2   DOF 2: "Men are dangerous"

Alicia continues to discuss her claims while her classmates present new angles on the same issue. The teacher interrupts and ask about the male role: What will happen if men are described as constantly horny and without control? What consequences arise from such a generalisation?

Alicia: Then, it's individual how people are, so, it's sort of difficult. You shouldn't generalise, but it's hard to do something about it just because it has become kind of a norm … that if a girl walks home by herself and happens to see a boy she becomes a bit scared, even though he is the world's kindest. It's unconsciously generalised that men are dangerous. It's hard to do anything about it just because …

Anna: Because it's like the norm. That guys are always so damn horny. It's also that you are lumping all men together …

Lotta: Isn't there some percentage … that rapists … that it's only men … that it's kind of 99% men. Then it's no wonder you think so, but it is still wrong.

Fredrika: I think if you want to have sex with a girl and the girl doesn't want to, it's still possible … you know, vaginal sex. If a girl wants sex with a guy, it will not work unless the guy has a hardon. It rests very much on the guy, I think.

Monica: Because men can't be raped?

Lotta: So, of course they can. But I mean that they can't be raped by vaginal sex. If they don't get a hardon, they cannot …

Monica: So, there are other ways? We don't have to go into details … It is possible to abuse. But it's incredibly hard to think that … It's likely that the statistics are right, that there are more men. But what would happen if a man was raped by a woman?

Ida: He's looked upon as rather weak.

Sandra: It's like such a tremendously hard norm to kill. I don't even know where to start.

Now the conversation leads to a new norm about girls' vulnerability, about how girls are expected to be afraid of boys due to the danger of their strong sexual desire. But also, about how the men who fail to live up to this norm are regarded as weak. The students also discuss how rape statistics fuel the norm about men being dangerous. In Table 6.2 we present a compilation of the ethical questions that the students raised in their discussions, taken from DOFs 1 and 2.

6.3.2   Example 2

6.3.2.1   DOF 1: "The length of the act of penetration is important"

In another group (mentioned in the introduction as having discussed the norm of penetration) Viktor shifts focus, introducing the idea that people generally believe penetration should last for a long time, (DOF 1). Another student in the group, Ulrika, agrees:

Viktor: Do you only count penetrative sex or do you count the whole action? Or, homosexual … namely gay sex, or as between several partners … so, what is the range? And how do you regard sex in general — is it just the orgasm that has to be the goal? Or can sex just be a little … can sex just be a little … that you just touch each other but nothing more and …?

Ulrika: Yes, where are the boundaries in relation to …? Where is the limit of calling it sex?

Viktor: Yes, exactly! I think a lot comes from the porn industry and that like … it should be … for sex to be … A guy should be able to keep on going for a long time without coming and thus be able to give a woman pleasure. She should come several times and it should be good, according to the norms.

In the situation, Viktor has chosen to raise some issues which he looks upon as a common norm in this context, i.e. that the sex act should last for a long time. He continues, "it's supposed to be an intense fuck for at least half an hour or so to make it count as good".

6.3.2.2   DOF 2: "Sex should last for a long time"

When studying what is happening in the conversation it becomes clear that the student discussions lead the norm-critical analysis further than the students were required to go. The discussions raise several new norms they need to consider. First, they engage the question of what should really be counted as sex. Is there a norm in our society that prescribes what counts as sex? The conversation continues when Ulrika addresses what Viktor said about the impact of porn on norms.

Ulrika: I thought it was interesting what you said about porn. The porn industry apparently comprises 90% men. It is created by men.

Viktor: Yes, exactly.

Ulrika: And it is men who have created the norm that they should have a big penis, and it feels a bit like it's about the same thing, because it is maintained. Porn has a big impact and it is maintained by men who say that the sex act should last for a long time.

Viktor: Yes, if two [men] are talking, then maybe one says, "I had sex for an hour, and I made my partner come many times" and "it was so good". Then the other, maybe, goes like, "Oh, I have to beat that".

According to Viktor, the length of the sex act becomes, in this way, a norm that contributes to competition between men. An ethical issue raised here concerns the question of whether it is the men who determine the norms.

6.3.2.3   DOF 2: "A woman should come through vaginal sex"

Now the students have distinguished a new norm in their conversation — that intercourse should last for a long time, which quickly changes to another norm that is about the size of the man's penis.

Viktor: It's the same as "I have a huge penis".

Ulrika: Yes. "I want", "I have a bigger one". But actually, most orgasms don't come from penetration, it is the clitoris that … So, that's quite interesting … the length should not really affect …

Viktor: No.

Ulrika: … if the orgasm mostly comes from the clitoris so to speak.

Viktor: Must it then be … Usually, only vaginal sex counts as sex … the norm is that a woman should come through vaginal sex.

Here another two norms are apparent. One is that sex should mean vaginal penetration, and the other that this is framed by heterosexuality. The ethical issues discerned by the student group in example 2 are summarised in Table 6.3.

6.3.3   Example 3

6.3.3.1   DOF 1: "It's going to hurt and will bleed"

In another group, Sofia has investigated what she has perceived as a norm, namely that women should bleed at first intercourse and that it should hurt (DOF 1). This is a norm that she believes is still predominant in several cultures. The ethical question raised here relates to norms about the female body.

Sofia: The statement I wanted to investigate was if it is true that all girls bleed the first time they have vaginal intercourse. I have chosen this because it was something I believed, not so many years ago when I was in high school — that one actually should bleed the first time and that it should hurt. That was how it should be. And this is associated with this myth of the maidenhead — that you have a membrane covering the entire opening, and it will burst and then you will bleed the first time. But there is no such thing. Instead, you have a skin fold, a better word for this is a vaginal corona. So, maidenhead is perhaps a word we shouldn't use at all. And this, about bleeding the first time, it has caused a lot of problems for a long time. For some, it has really become important to bleed the first time … in some cultures and religions. For others, it's been, like, somewhat scary, the bleeding and the pain. That it has to be like that. I watched a programme where they interviewed a doctor about this. He practices a surgery there … for girls who worry about not bleeding the first time.

Klara: Are they worried about not bleeding?

Sophia: Yes. Girls from such cultures where it is very important for them to bleed the first time, they … in order to prove that you're a virgin; to prove that this membrane, which does not exist, is there …

Klara: Mm [agrees].

Here, Sofia points out the myth of the so-called maidenhead, which is expected to burst when girls/women have intercourse for the first time. First, she gives a scientific, anatomical description of the myth of the vaginal corona. Then she talks about the expectations and concerns that are associated with this wrongly described 'membrane'. Then she goes on to say that there are doctors in Sweden who perform surgical procedures to help vulnerable women who want to prove their virginity.

Sofia: And then there are two ways to do it. Either way you insert stitches that will cause bleeding. Or you sew these skin folds you have, so they become like a membrane. And I reacted quite strongly to the fact that, like, here in Sweden, one can go and get such surgery. And it's like this … that way you can help … or keep this myth alive. So, I think it's really important to help girls from such cultures where it's important to bleed the first time. But creating such a membrane may not be the right way to help them. And this doctor also advises them on how they can make it seem as if the membrane burst. That they should hurt themselves. That they put something sharp in bed, so … It was just like this …

Sofie points to an ethical dilemma where the act of helping women simultaneously helps to maintain the myth of the maidenhead. The discussion between the students continues to address the problem of how menstrual bleeding takes place if there is supposed to be a membrane across the vaginal orifice.

Berit: What about when having a period?

Sofia: Yes, so if there was such a barrier, it would not have been possible for menstrual blood to run out or discharge. So, it's, like, completely …

Klara: But wouldn't it be better then, for this woman … or these women who are afraid that it will not bleed, that they could bring their men to the hospital, so that the doctor can explain.

Sofia: Yes, I think so. Such a solution would have been a better …

Ready: Yes.

Sofia: … rather than make it look as if this membrane exists. Because then this myth continues to live and …

6.3.3.2 DOF 2: "It's important to show that you are a virgin"

The conversation raises a new ethical dilemma: why not rather put effort into informing men of the scientific fact that there is actually nothing that bursts during first intercourse. Sofia states that the dilemma is linked to how you regard intercourse in other parts of the world. Klara is wondering if this kind of surgical procedure is performed openly in Sweden or if it is done under cover.

Sofia: In Sweden … it is still quite modern and so on. That one can still do this, I think … [depends on] how it is in other parts of the world.
Klara: But these operations, are they done at regular hospitals or are they done under cover, in secrecy?
Berit: No, I think they …
Sofia: It is private clinics that do them, I think.
Berit: But is it wise … private clinic, or are they dodgy?
Sofia: Oh yes …
Klara: I think like this, academy … clinics that are well known. That they would do such surgery …
Berit: No. Although they might think that it is important for the woman, that she should decide for herself … if she really wants it.
Walter: Precisely, yes.
Klara: I had no idea that you could perform such surgeries in Sweden.
Walter: I've heard about this before, that it's important to show that you are a virgin and about that … myth. But never about the fact that some act to fix it.

In this discussion, the students revealed additional ethical questions. Is it okay to perform this type of surgery, and who actually performs the procedures? What is most important — to satisfy the women's perceived need to be operated on or to choose not to operate because it's really just a myth? Eventually the conversation moves on to discussions about how the fear women experience can affect them to the point that they fail to become aroused at all and how that in turn can lead to bleeding.

Pauline: But it's not … they're not just doing that to show you are a virgin. It also has to do with … to justify that it's hurting the first time.
Sofia: Yes, there are girls who are restricted because of that. That they will be scared and just say, "no, it will hurt, I will bleed, I do not want to".
Pauline: And it's kind of normal, that it's going to hurt.
Sofia: Yes, it's going to hurt.
Pauline: It depends … then something is a bit wrong.
Berit: Yes, maybe she is not aroused then.
Pauline: Or, if she's nervous then it can be so.
Sofia: But there are those who bleed. Isn't it about 30% who bleed the first time? And it's not because there's some membrane that bursts. It is because you are not aroused enough or that you are too tense.

Pauline: Or for some … vaginal corona … can …
Sofia: Yes, it can.
Pauline: It may break a little.
Sofia: But there is nothing that breaks or disappears or something.
Klara: But you cannot see if someone is a virgin.
Sofia: No.
Pauline: But well, it's kind of that I've got rid of it now, I can never be a virgin again. You're never completely clean.
Sofia: But there's nothing "to get rid of".
Pauline: That's what I mean, there's nothing to get rid of. But you kind of try to make it as if it was so … that's why you have to wait, because you'll get rid of it.

What we have seen here is that when students get the opportunity to discuss the 'taken-for-granted' norms, it creates space for a series of new ethical questions. The ethics identified in DOF 1 lead to new areas in DOF 2. For example, we can see how Sofia's norm investigation about bleeding as evidence of a woman being a virgin or not led to a number of other issues to consider — issues to stand for or against. Should the patriarchal structures found here be challenged, and should the men who perpetuate them be informed so that they learn that there is no covering membrane. Or should the cultural tradition survive? Should Swedish society protect scientific findings, or should other customs and experiences be allowed to fit within the framework of Swedish society? What significance should knowledge about biology, physiology and anatomy be allowed to play in that discussion? In Table 6.4 we summarise the ethical concerns that arose in example 3.

6.3.4 Concluding DOF 1 and DOF 2

In the classroom interaction reported above we were able to follow groups of students in discussions which departed from their own norm-critical investigations (DOF 1). Furthermore, based on their intuitive feelings which arose in a narrow conversation, our analysis showed that students were able to identify a diverse range of new ethical issues (DOF 2). What an analysis of these conversations further shows is that the students conducted a discussion that revealed a variety of ethical dilemmas and positions that would not have been raised outside of such a communicative exchange. It is obvious that the teaching methodology gave the students opportunities to tackle these issues from a variety of angles, raising numerous interesting ethical questions (summarised in Tables 6.2, 6.3 and 6.4). Through their resistance to excisting norms and a critical review of them, a new flow of content where meaning constantly shifted was raised (Lenz Taguchi 2004). The creation of learning spaces where students are given such opportunities should be a basic design principle in all education where science intersects with ethically loaded content.

6.4 Researchers' Steps Backward into the Depth of Society — A Question of Responsibility, DOF 3

In this final section we emphasise that all teaching flows from more or less conscious choices based on the curriculum and practical reality where certain content is foregrounded and other backgrounded. Following Foucault's (2002) writings we reflect on the students' discourses in relation to the historical and social context in which they are embedded (DOF 3), and discuss the implications for research ethics.

We begin by giving a brief 'genealogical' background to the human search for ultimate reason. Many philosophical reflections (c.f. Dewey 1929), declare that early in human history, during a lengthy animistic era, we created structures, essences and entities in nature to describe the origin of emerging phenomena. Soon those explanations became refuges and safe places to rely on when nature appeared in its most insecure guises. Later, when the western tradition (with its origins largely in Ancient philosophy, and then Christianity) took over, more generalised metaphysical representations concerning the order of existence came to have wider expression. The answers to life's big questions, which people sought primarily in universal principles of nature and social life, thus became dependable superordinate principles to guide people through the immediate struggles of everyday life (Dewey 1929). Subsequently, from these ideal principles, power relations and social hierarchies such as family formations, ethnicity, roles in trade and economics, as well as class and gender, were further crystallised to form permanent power structures in western culture (Honneth 2008).

However, during The Enlightenment people started challenging such predominant systems. Increasingly the answers came to be informed by science rather than religion. Thus, science, and above all, biology, became an alternative paradigm in the pursuit of identifying ethical maxims. Contrary to searching for rules given by divine power, one now rather asks for what can be identified as 'natural' based on science, biology and evolution. Accordingly, in teaching about sex, sexuality and relationships, biology has assumed a particular position as a basis for what is considered 'natural' in relation to questions about the human body and human behaviour (Barron & Brown 2012).

In this chapter, we followed a class that, in a non-confessional environment, had the task of examining prevailing views of sexuality in society. Based on a biological framework they asked questions such as: What is it in our common cultural assumptions that sets the limits for how we allow ourselves to think and act sexually? Should we take heterosexuality for granted? Are our gender roles biologically determined? Is orgasm the ultimate proof of desire and pleasure? Should we expect all women to bleed during first intercourse? Thus, the question we must ask ourselves as researchers is: What inclusions, exclusions and systems of power are inherent in these particular systems of thinking (Gytz Olesen et al. 2004)? In what way does this biological framework help us to challenge assumptions about what is regarded as natural human sexuality?

Certainly, what is classified as biological can be regarded as 'natural', but we cannot allow biology per se to determine human social behaviour. Gang rape among animals and the practice of male animals killing the offspring of their rivals to ensure the dominance of their own genes can hardly be regarded has healthy models of behaviour for the human species. What biology regards as natural can never be formulated as a role model in a human community (Orlander 2016). Certain delusions, manifested in brown coats, have already tried this the spirit of social-Darwinism (Crook 2007). Moreover, contemporary researchers have shown how some 'objective facts' produced in the natural sciences are often pervaded by a social ideology where notions of sexuality and gender have a significant impact on how the biological content is interpreted and presented (Ah-King et al. 2014). Historically the notion that human being is 'naturally' hetero-sexual has been extrapolated from sexuality as it is described in research on animals. It is better to define the criteria for human sex and interrelationship on the basis of human values and deliberation. Within this context, the biological perspectives of course create an important resource among many others. Furthermore, it becomes apparent that the students contribute to broaden the view of what is to be counted as a relevant content in the school-subject biology. However, as researchers (and perhaps also as teachers), we may need to take a step back in relation to the whole teaching situation. Based on a genealogical framework (Foucault 2002), one can grasp the idea of the historical and cultural scaffolding within which teaching takes place (DOF 3). What is considered natural within a biological framework may not always be applicable to the human context. In summary, the third depth of field, DOF 3, focuses on what becomes visible in the transaction between what is happening in the classroom and the historical and social context within which this takes place (Foucault 2002), and what then, from a research ethics perspective, becomes important to take into account. However, the position one chooses to take on this issue is ultimately a question of what consequences we are willing to take responsibility for as a researcher, and this is what is highlighted from the position of DOF 3.

References

Ah-King, M., Barron, A. B., & Herberstein, M. E. (2014). Genital evolution: Why are females still understudied? *PLoS Biology, 12*(5), e1001851.

Barron, A. B., & Brown, M. J. F. (2012). Science journalism: let's talk about sex. *Nature, 488*, 151–152. https://doi.org/10.1038/488151a.

Crook, P. (2007). *Darwins coat tails. Essays on social Darwinism*. New York: Peter Lang publishing.

Dewey, J. (1929). *The quest for certainty: A study o the relation of knowledge and action*. Montana: Kessinger Publishing.

Dewey, J. (1957). *Human nature and conduct*. New York: The Modern Library, cop.

Dewey, J., & Bentley, A. F. (1960). *Knowing and the known* (p. 1960). Boston: Beacon Press.

Foucault, M. (2002). *Vetandets arkeologi* [Archaeology of Knowledge] (C. G. Bjurström & S.-E. Torhell, Trans. [Ny utg.] /ed.). Lund: Arkiv.

Gytz Olesen, S., Møller Pedersen, P., & Johansson, I. (2004). *Pedagogik i ett sociologiskt perspektiv: en presentation av* [Pedagogy in a sociological perspective: a presentation of]: *Karl Marx & Friedrich Engels, Émile Durkheim, Michel Focault, Niklas Luhmann, Pierre Bourdieu, Jürgen Habermas, Thomas Ziehe, Anthony Giddens.* Lund: Studentlitteratur.

Haraway, D. (1988). Situated knowledges: The science question in feminism and the privilege of partial perspective. *Feminist Studies, 14*(3), 575–599.

Honneth, A. (2008). *Reification: New look at an old idea.* New York: Oxford University Press.

Lenz Taguchi, H. (2004). *In på bara benet: en introduktion till feministisk poststrukturalism* [Down into bare bone. An introduction to feminist poststructuralism]. Stockholm: HLS förlag.

Orlander Arvola, A. (2016). So what do men and women want? Is it any different what animals want? *Research in Science Education, 46*(6), 811–829.

Östman, L. (1998). How companion meanings are expressed by science education discourse. In D. A. Roberts & L. Östman (Eds.), *Problems of meaning in science curriculum* (p. xii, 288). New York: Teacher College Press.

Rogoff, B. (1995). Observing sociocultural activity of three planes: Participatory appropriation, guided participation, and apprenticeship. In J. Wertsch, P. Del Rio, & A. Alvarez (Eds.), *Sociocultural studies of mind* (pp. 139–164). Cambridge: Cambridge University Press.

Swedish Research Council. (2017). *Good research practice.* Stockholm: CM-Gruppen AB.

Auli Arvola Orlander has a great part of her professional life been involved in various issues in the school world. With a background as a teacher and later as a teacher educator, educational consultant, etc. she has been in contact with practice-related issues in the field. She has been involved in various research and development projects within the school and nursery, all in close collaboration with principals and practicing teachers. Today Auli is working in the Department of Mathematics and Science Education at Stockholm University as director of studies. Her main research focus is on gender issues in science education.

Iann Lundegård is an associate professor in Science Education at the department of Mathematics and Science Education at Stockholm University and partly employed by SWEDEST at Uppsala University. His research interest is educational philosophy associated with high school students' deliberations and meaning making on sustainable development. He has quite a long experience in working with pre-service and in-service teacher training in Science Education and has written several textbooks aimed to be used at upper secondary school. Over the years he has contributed to curriculum development at the Swedish Agency of Education, and recently on behalf of them, also produced texts and other material that may support teachers for didactic reflection on their teaching in/on/about sustainable development.

Chapter 7
Challenging Existing Norms and Practices: Ethical Thinking at the Science Education research Boundaries

Jaume Ametller

7.1 Introduction

The five chapters included in this section present different ethical issues that concern science education research. The issues are varied and so are the ethical problems they pose and the solutions that the authors put forward to address them or, at least, to reflect upon them. My aim is not to discuss or summarise them but to suggest common elements which might be useful to help to interpret and to contextualise ethical research problems in science education in a way that is relevant to the new challenges these five chapters exemplify. In doing so, I will lose the depth and richness associated with the particular cases developed in the previous chapters but my goal is to abstract from that richness insights that might be useful in a variety of situations, including, but hopefully going beyond the particular examples of this section.

I will start by attempting to define the nature of these challenges in terms of where they take place. Once I have discussed their nature, I will present some possible sources of those challenges in terms of some constitutive aspects of science education research. Finally, I will propose ways to address these challenges by considering both how we could conceptualise and reflect upon ethical issues in science education research and how we could guide our ethical decisions. Through the chapter, I will advocate for centring our discussion about research ethics on a particular field, that of science education, so that the challenges I discuss, and the ways of engaging with them, will reflect the nature of the network of actors that take part in it (Latour 2007).

J. Ametller (✉)
Department of Specific Didactics, University of Girona,
Girona, Catalonia, Spain
e-mail: jaume.ametller@udg.edu

© Springer Nature Switzerland AG 2020
K. Otrel-Cass et al. (eds.), *Examining Ethics in Contemporary Science Education Research*, Cultural Studies of Science Education 20,
https://doi.org/10.1007/978-3-030-50921-7_7

7.2 The Nature of the Challenges: Individual, Social, and Content Domains

The chapters in this part of the book present a variety of ethical challenges which are representative of the issues encountered in a field as diverse as science education research. Despite this diversity, the issues show some common themes that talk to the shifts in interests and approaches to research in this area. I have grouped these challenges around three foci of ethical issues for presentation purposes but in many cases the research situations in which ethical issues arise involve the interaction of more than one of these foci: issues connected to the social responsibility of research, issues located in the interaction between researchers and participants, and issues connected to the ethical elements of the content being taught.

7.2.1 Issues Located in the Protection of Participants Rights – Individual Ethics

This is the area most commonly discussed in research ethics: how can we ensure that participation in research will not harm participants. Educational research will very rarely entail the risk of physical harm to participants but other types of damage need to be considered, such as those related with participants social exposure (Burbules 2009) and the personal investment on time and effort versus their gain (British Educational Research Association 2018). In the case of education, some of the participants are particularly vulnerable both because they might be children and because research might be focusing on particular groups already socially vulnerable. In any case, researchers must reflect on how participation could put them at risk, and conduct research accordingly to eliminate or minimise such risks.

Once the aims of the research have been established, realizing them is considered by those involved as attaining something worthwhile or beneficial in some respect. Despite this desirability, the process to accomplish those aims must take into account ethical commitments towards the wellbeing of those participating in the research. Research ethics has placed not harming participants beyond any consideration of gains that might be obtained through the research. However, as we have seen so far, and as I will discuss further in Sect. 7.4, the definition of harm in science education research rarely includes physical injury and, in many cases, it has more to do with the idea of respect or avoiding being detrimental in ways that might place researchers in the situation of legitimately wondering if the "greater good" of the expected results does outweigh the disturbance participants might experience. Such choices might entail deciding if informed consent can be restricted so that participants' knowledge does not negatively impact on the research, or whether to adopt an experimental methodology with control groups-even when BERA's ethical code (BERA 2018) suggests otherwise-because the alternative methodologies won't provide results with enough weight to change policy. Issues of this kind have to take into account socio-political changes, such as the strengthening of the accountability culture, and socio-cultural changes on how identity and agency are constructed and enacted and, hence, how this affects what must be understood by harm and to what extent individual participants can make decisions on these issues.

Issues connected with the impact of research on participants are often connected to the research methods being used and the relationship between researchers and participants (chapter by Allison and Vogt this volume). Measures taken often have to do with methodology, but in the cases where some kind of relationship is established as it is often the case in qualitative or mixed methods research, "non-methodological engagements", i.e. situations of interaction outside the commonly understood and agreed research situations, must also be carefully considered. Partly the problem, as we will discuss, has to do with how researchers and participants see their interaction in different moments as wither part of the research activity, hence subject to research agreements, or as part of a personal relationship ruled by other codes of conduct, which might or might not be shared, but which have usually not been discussed or agreed upon.

Research methods involving video data, information from social media, or internet mining will likely collect information connected with how "identity" is socially defined now. These technologies have a particular impact on issues connected to personal interaction. Visual methodologies are hardly a novelty in science education research, but visual data is more easily collected, edited, and distributed; this ease has fundamentally changed the relation of individuals with video data of themselves (Ametller 2008; Derry et al. 2010). The relation of participants with data, including visual information on themselves, have changed in the past years when this media has been widely socialised as a way to express and construct personal identities (Adami and Jewitt 2016). Ethical questions around anonymity and privacy are part of wide social discussions which reframe the research ethics discussion around these issues. On the one hand, measures to be taken might have to be more stringent because of how easily this data can be distributed but, on the other hand, participants are likely to be more informed about the use of images and to have more agency on the use of their images, which might entail repositioning the researcher's role from one of protector to a more equal standing, one more akin to negotiation than to overprotection.

Networked technologies are more novel than audio-visual technology in educational research and some of the issues they generate – for example, the consideration of anonymity in contexts where participants are users of social media (British Educational Research Association 2018; Burbules 2009) – are not yet well understood and appropriate, normative or accepted ways to address them are not established. Some of these issues are connected with the new uses of visual data mentioned before but also create other challenging situations. In a hypercultural society (Han 2018) the presence in the network of both individuals and institutions can be a source or research data and also an important part of someone's identity. This generates new issues related to what can be used as data and how to obtain permission and because interest on how learning happens in the continuum of time and space (including hyperspace) those are issues that will likely become more prominent in ethical discussions. Networked technologies also make social networks relevant when considering the impact of participation in research and how this can be perceived outside the group of direct participants (chapter by Ryu this volume).

7.2.2 Issues Connected to the Social Responsibility of Research – Social Ethics

Science education research does not only concern those directly involved in a particular study. It is a social activity with a structure intertwined with other social actors: schools, universities, education policy-makers, funding bodies…. These actors influence science education research practices in different ways. Some of them are aiming at the research activity per se (defining aims, restricting access, directing funding, etc.), others affect research more indirectly (university appointment procedures, educational policies, curriculum reforms, teacher training programmes, etc.). As a whole, they constitute a network of multiple agencies that shape science education research. Negotiating this network also involves ethical issues (see chapter by Johansen and Anker this volume). While the above section has focused on ethical issues more directly connected to individuals, this part looks at issues primarily connected to the social dimension of science education research and its relevance (chapter by Gimmler this volume). I will focus here on the relationship between social responsibility and care for individual participants when deciding to what extent the greater good of research results might justify some aspects of how research is conducted.

Over the past few years, a fundamental issue for social sciences researchers has been the accountability of their work. Mostly, this means being accountable to those who are funding and or supporting our research, either directly or indirectly. This implies that there can be an economic measurement of the value of research which then justifies the influence of those providing the funding on what is being researched and, in some situations, what is reported about it, when and how. These issues affect an important part of research ethics, i.e. the independence of researchers. Ethical problems arise from the fact that extreme positions on this question would either undermine the credibility of the research undertaken or make it impractical to pursue. Researchers must retain a degree of independence but recognise, both for themselves and when sharing their work, the influences that have affected them in their work. Such influences do not need to be seen as a lesser evil, but actually as a normal consequence of being part of a network which makes it possible to affect others in the network, i.e. to make it possible for research to have an impact on teaching practices. Being part of the network also means, that there will be tensions, for instance, around the aims of a particular activity (Engeström 2015). The negotiation on the aims of the research starts even before contacting possible participants. Are these aims included in the funding programmes available? Will these aims lead to outputs that are perceived as valuable by employers (i.e. will I get a job or a promotion at a university because of the particular research I have conducted)? Should the aims be aligned with policy or should it be connected to issues perceived as valuable by schools? How much influence should one allow to make a relevant contribution while retaining independent judgement? How much control over the results must one retain? Should aims be negotiated with participants? If so, where is the balance between respecting agency and upholding one's professionalism?

7.2.3 Issues Connected to the Nature of What Is Being Taught in Science Education – Content Ethics

Contents touching upon ethical aspects of science as an activity and of science as part of social debates with ethical ramifications have become more present in schools due to the orientation of curricula in recent decades. While this is not research ethics per se, it touches upon the two previous types of issues. On the one hand, because researching in these contexts is likely to face personal issues with ethical ramifications. On the other hand, because the research methodologies and, more widely, how researchers participate in the activity and their interaction with participants need to be considered including those ethical issues.

Beyond the ethical side of chosen scientific topics (Jones et al. 2010), the choice of topics related to personal identity or believes such as religion (Reiss 2008), or that has to do with students' behaviours such as sexual relationships (chapter by Orlander and Lundegård this volume) might place ethics at the centre of the reflection on the wider social responsibility of science education researchers. In both accounts, researchers might face ethical issues connected to cultural and believes diversity in the classroom.

The three dimensions suggested in this section are similar to those encountered elsewhere in the literature in terms of social responsibility and individual protection (Tangen 2014) but has chosen to explicitly acknowledge the content being taught as a dimension. This choice is motivated by the increasing importance of competences and contextualisation in science education curricula which, coupled with a grooving interest on identity issues, is likely to bring to the fore ethical ramifications of contents being addressed in science teaching. On the other hand, the section has not presented a dimension connected to the researchers and the research community. These aspects will be addressed in the next section as one of the two source of current ethical challenges.

7.3 Sources of Ethical Challenges

In the previous section, I have presented the loci of ethical issues that authors in earlier chapters of this book have noted as being relevant in their science education research, and which could hopefully be useful to locate other science education research. Why are these issues relevant in this area of research now? I propose in this section two possible answers. On the one hand, *onto-epistemic issues* of science education research, on the other hand, *intrinsic issues* of the field and those working on them, their boundary crossing and boundary actor characteristics. I believe these two sources will help to explain the issues presented in this book but could also be of help to other science education researchers to work with or address these issues.

7.3.1 Ethical Issues Connected to Onto-epistemological Choices

Concepts such as identity (personal, social, religious, technical, national *etc*), and agency and values are important ontologies. The socio-cultural influence in educational research means that these concepts come to the fore in a wide range of research work. Even when these concepts might not be the focus of the specific research, they are likely to be elements relevant to the theoretical framework even when these concepts might be defined differently in different frameworks because these are ontological elements that speak to the fundamental point that education is a social phenomenon. Therefore, while these terms might be associated in educational research, if loosely, to post-modernist philosophical takes they would also be relevant in, for instance, new-realist approaches where aspects such as identity or agency can be seen as emergent characteristics of a particular social assemblage (DeLanda 2006).

These ontologies are linked to the focus on social interaction in educational and sociological theories that are prevalent in the field. This theoretical focus is connected with a wide use of research methodologies that require the personal interactions between researchers and participants which, as we have seen in the previous section, involve situations that might entail ethical challenges. Therefore, ontological commitments might introduce in our research ethically sensitive elements that are connected to epistemological choices, and epistemological choices are connected to methodological practices which bring about situations that might entail ethical challenges. For instance, the inclusion of identity as an ontology, even if it is not the focus of the research, is likely to be associated with an interpretative framework of science education teaching and learning as an activity which will take into account the relationships between individuals in social contexts. These interpretative frameworks are likely to be associated with data, and the methods used to collect it, which might introduce in our research practice information deemed sensitive. Even if we don't want it, we have "asked" for it, we have it now, and is a part of our broader interpretation and, hence, we need to deal with it ethically. The choices we make on the ontologies we consider relevant shapes how we intervene, engage and interact with the sociomaterial system we are researching in a way that we are *diffracting* (Barad 2007) the information from that system, showing some aspects of it and not others, because of our particular intervention. We are ethically responsible for both what we choose that matters and what we exclude from our account.

While onto-epistemological elements provide insights on the topics around which research ethics issues will arise, to better understand the nature of the ethical challenges faced by research in science education as an activity, we need to understand the network that gives rise to it. In the next section, I will focus on how the network of social communities and practices shape the characteristics and practices of science education researchers and how these can help us understand some of the ethical challenges we have discussed so far.

7.3.2   Science Education Researcher as a Boundary Actor

Science education research is often described as an interdisciplinary field (Sjøberg 2007) but I would argue it often resembles more of a frontier difficult to inhabit because it stands between areas with little or no overlap. From its foundations, science education research has searched for inspirations in both the science camp and the psychological and education camp. While the field has been successful at generating humble theories (Cobb et al. 2003) that provide useful constructs to understand the process of teaching and learning science, it has not generated a theoretical framework that brings together the original sources of those fields and this is probably as it should be because these are fields with important ontological and epistemological differences. The result is that knowledge and values from areas with little overlap are part of the research and practices in science education. The literature on boundary crossing and boundary objects provides a useful framework to explore this (Akkerman and Bakker 2011).

From its definition by Star and Griesemer (1989), boundary objects are information or objects used differently by different social communities. Aside from boundary objects, we can also consider boundary actors which are "politically motivated actors who manipulate social processes across communities and whose reflexive actions inhibit boundary objects individuals (Star and Griesemer 1989). We use the concept "boundary actor" to denote the individuals who mediate between incommensurable paradigms in the context of power inequalities (Keshet et al. 2013, p. 668). I suggest that science education researchers can be seen as boundary actors operating in a network of relations among school, university, and often policymakers, communities.

On top of the boundary aspects of the practice of science education research, science education researchers are often professionals with a double background, in experimental sciences, in social science research and, in many cases, in teaching. This diverse background is part of their identity, a multifaceted identity that means that researchers might find themselves in contradictory roles in research settings. It is common to play the card of being an educator to gain access, but that might be at odds with the aims as a researcher. This multifaceted identity does not only have the potential to generate challenging situations with participants but also self-conflicting situations for the researchers. The fact that this boundary identity is at the heart of some of our ethical problems is probably not unique to science education but is probably best addressed by placing this ontological characteristic at the basis of our ethical reasoning.

I want to stress that science education researchers as boundary actors do not only mediate among "external" communities but also mediate among their own "internal" communities. They have boundary identities which allow them to be boundary actors but often those "boundary identities" are conflicting and instead of creating a "mixed identity" generate multifaceted identities that can show individual sides to external actors. The connection of the transaction of those "mono-faceted" presentations with the rest of the (necessary) internal identities generate research ethics problems related to how researchers present themselves to each of the social communities they are part of, or interact with, as well as ethical problems connected to the researchers' decisions on the work they choose to do which might hold different value in different communities.

Often the challenges connected with frontier characteristics of science education research and researchers are also connected to power relations established in research settings. There are several ways in which these power relationships can manifest in science education research for instance on gaining access to participants, practice settings and data. A particular relevant way in which power relations can relate to ethical issues is the establishment of the research aims. Research funders, university policies, and political priorities play a role on the definition of these aims and are all part of the boundary character of the researcher as an actor but for our discussion, we will focus on how they are dealt with by researchers and participants. This involves, at a basic level, the need for informed consent but it often involves some sort of negotiation with the participants in co-constructing some aspects of the research. In doing so, questions of power over the activity and of whom the research will benefit are negotiated.

We can find in the literature (see chapter by Ryu this volume) specific examples of how ethical challenges generated by conflicting aims in the co-occurring practices on science teaching and learning and science education research were addressed through different ways of sharing power. Most of the time this involves negotiating with participants some decisions to do with the planning, conducting and reporting of the research. This runs several risks and, finally, is problematic because those are solutions that, as the discipline itself, often try to find common ground where there isn't and hence the solutions fail at bringing together different perspectives, aims, knowledge or values and, instead, the outcome is usually a de-professionalization of researchers or a shift towards innovation rather than research. The former might happen if the researcher decided to address ethical challenges by reaching a consensus with the participants on how the research will be conducted instead of assuming the role of the expert in research. The later can be a consequence of constraining research designs to ensure that participants will benefit from taking part in the research. I see the reticence to use experimental designs with control groups as an example of this which might constrain the scope of more fundamental research.

7.4 Addressing the Challenge: Ethical Thinking and Ethical Decisions

In this section I will present a proposal for addressing these challenges to provide ideas to be applied generally to the discussion of ethical issues in science education research and how to address them. The proposal is aligned with existing ethical codes, BERA (2018) for instance, which advocate for having some guidelines but ultimately considering each case on its own merits and specificities, hence moving on the direction of pragmatic or virtue ethics. However, I will try to make some specific suggestions that are intended to respond to the issues I have discussed in sections 2 and 3.

In the previous section, I have presented two different sources for ethical challenges but it is worthwhile noting that they are often related and that they are not easily dismissed. By pointing out the sources of challenges, and sometimes the reason why those challenges are difficult to resolve, I am trying to show that making ethical decisions as science education researchers will require considering these issues, frontiers and power relations, and while deontological solutions are an unlikely option given the complexity and contextual influence of the situations involved, there is a need for generating guidelines which are centred on the characteristics of science education research.

While the community of education research has produced several ethical codes and guidelines that are widely acknowledged it is not frequent to find research ethics discussed in depth in research papers in science education. This does not necessarily mean that the available guidelines provide all the required answers. A previous chapter in this book (chapter by Allison and Vogt this volume), for instance, discusses how commonly used deontological guidelines, originally based on bioscience research, are not adequate to address educational research ethical challenges. The rest of the chapters do not explicitly reject existing guidelines but do not directly derive their solutions to ethical challenges from them, rather they provide particular solutions based on their reflections which might, or might not, be explicitly referred to the literature.

To shape our ethical reasoning according to the needs, situations and challenges of science education research we should consider more explicitly the foundations of our ontological and epistemological thinking. This will help us to gain clarity on what are the objects of interest of our research and how we consider we might construct valid knowledge about them. What degree of anonymity, for instance, will prevent us from studying what we need to study?

I have argued before that identity and agency are important ontologies in science education research now. Issues of identity are culturally defined and valued. In hyperreality (Han 2018) this has moved towards a more delocalised, *deterritorialised* in DeLanda's terms (DeLanda 2006), which is accompanied by a drive towards globalisation or increased coherence across our species. This drive is, however, accompanied but another one that moves towards differentiated, person-alised unique identities which are constructed in a self-aware, purposeful way, through a greater sense of agency while also oriented towards social validation. This is to say that while participating in a potentially global network of relations defining identity, each individual is more likely to be different from those close to him or her. This double movement is coherent with the idea of having general guidelines but focusing on contextual discussions since it suggests that the globalisation vector can provide guidelines on the issues we need to consider while the particularised move-ment is pointing us towards taking also the agency of participants into consideration when facing particular situations.

This way of proceeding might have consequences on how we approach some of the most commonly considered ethical issues. For instance, we might wonder if ways of preserving privacy make sense now when moving from deontology to personal agreement with participants whose agency is seriously considered. Informed consent might also need some different approach. In a society adhering to democratic values we must question how far we need to agree on what we want to do "together" with participants. We risk a form of *enlightened despotism* if we are not prepared to enter a dialogic engagement which is compatible with maintaining different degrees of responsibility for particular decisions to do with the research.

A point of transformation of ethical thinking will be the determination of greater good and aim for science education research which is not shared across funders, actors and participants. I feel that once we reach agreements (local and contextual as well as more generic) on this issue we can move to power relations (and value/aim/methods) negotiations. If participants see their power of providing access in the framework of "greater good" and researchers see their role as experts as a way of securing social value to results the idea of defining worthwhile ethically admissible aims will be easier to agree upon.

Why do we want to reduce power gaps? It is a political aim? Isn't ethics dealing precisely with the existence of power differences where those with the upper hand must act in a way so that this difference in power is not misused? This entails that there might be a power relation that does not represent a misuse of that power. Maybe we should be looking for ways of empowering researchers and participants each in their areas of expertise or responsibility and work towards generating syner-gies. After all, several of the proposed solutions involve a dialogic perspective and we must remember that dialogue remains creative while differences are respected and maintained. Some authors suggest the use of boundary objects, for instance, a document of initial agreements on the aims and involvement of all actors, or the establishment of a protocol for periodically sharing perspectives on the ongoing activity, that can act as stabilisers of the tensions (Scoles 2018) by providing actors both a common ground of shared understanding and a way of channelling the ethi-cal issues that might appear during the development of the activity and that have not been foreseen in the original agreement.

How research is designed, carried out and reported has to consider the balance of respect for participants' identity and agency as well as for the social responsibility of researchers and the fields – negotiated in a complex dialogue among schools, policy-makers, funders, universities, etc. Keeping this balance, and taking into consideration social constructions of personal identity might push us to move from avoiding harm (a given in educational research) to respecting participants, and then to extend respect from participants to society at large – social responsibility. Is it in the balancing of the different foci of respect that we will encounter issues and responses? And this dialogical approach must be not just a measure of respect for the individuals taking part directly or indirectly, as well as for those who might benefit or be impacted by the research results, it must also be the respect for otherness and this changes our perception of our view and that of our field on science education.

In this section, so far, I have tried to show how engaging with the political and onto-epistemological ideas related to the ethical challenges we face in our research will provide us with guidance for ethical reasoning that is relevant and pertinent to the particular needs of science education research. It does help to reframe the situations we encounter to identify the ethically sensitive issues and elements so we can reflect and prepare guidance for researchers. Since I argue they can often be only guidance its application must be supported by something other than a deontological code. A possible candidate would be "virtue ethics". I am not claiming that science education research characteristics lead to virtue ethics but, rather, that this take on ethics fits with some of the demands that the challenges presented in this section, which can be generalised to a wide spectrum of possible ethically challenging research situations.

A consequence of this proposal would be to transform ethical committees into bodies that, through their discussions can play a role in training ethical researchers. This is important because to apply virtue ethics (Lovibond 2002) implies developing an ethical or moral character which can be, partially, accomplished through being exposed to ethical judgements. Furthermore, the ethical committees should look at border crossing in science education research, which would allow us to reflect upon ourselves as researchers and upon science education research itself.

To face the new ethical challenges, we encounter in science education research general guidelines will not be enough. We must be able to reflect, personally and collectively, upon the nature of those challenges, the issues underpinning them and the responsibility we have as researchers to participants and society at large. We need to see ourselves as entangled with different actors, both direct participants of the research and others who might benefit from it or that shape the research at different levels. This entanglement is not just a way of recognising the participation in a network but the understanding of how that participation changes all the actors, including ourselves, and carries a shared ethical responsibility (Barad 2007). If we do so, I believe that we will not only find ways of conducting our research ethically but, through that reflection, we will deepen our understanding of the field of science education research and increase its impact.

References

Adami, E., & Jewitt, C. (2016). Special issue: Social media and the visual. *Visual Communication, 15*(3), 263–270. https://doi.org/10.1177/1470357216644153.

Akkerman, S. F., & Bakker, A. (2011). Boundary crossing and boundary objects. *Review of Educational Research, 81*(2), 132–169. https://doi.org/10.3102/0034654311404435.

Allison, J., & Vogt, M. (this volume). Reflections on research ethics in historically oriented science education research in Canada. In K. Otrel-Cass, M. Andrée, & M. Ryu (Eds.), *Examining research ethics in contemporary science education research*. New York: Springer.

Ametller, J. (2008). Metodologías relacionadas con la utilización del video en la didáctica de las ciencias. Actas del XXIII Encuentros de Didáctica de las Cien-cias Experimentales. In: *Actas del XXIII Encuentros de Didáctica de las Ciencias Experimentales* (pp. 1270–1283).

Barad, K. (2007). *Meeting he universe halfway*. London: Duke University Press.

British Educational Research Association. (2018). *Ethical guidelines for educational research* (4th ed.). London: British Educational Research Association. Retrieved from https://www.bera.ac.uk/researchers-resources/publications/ethical-guidelines-for-educational-research-2018.

Burbules, N. C. (2009). Privacy and new technologies: The limits of traditional research ethics. In D. M. Mertens & P. E. Ginsberg (Eds.), *The handbook of social research ethics* (pp. 537–549). Thousand Oaks: SAGE. https://doi.org/10.4135/9781483348971.

Cobb, P., Confrey, J., diSessa, A., Lehrer, R., & Schauble, L. (2003). Design experiments in educational research. *Educational Researcher, 32*(1), 9–13. https://doi.org/10.3102/0013189X032001009.

DeLanda, M. (2006). *A new philosophy of society: Assemblage theory and social complexity*. London: Bloomsbury Publishing.

Derry, S. J., Pea, R. D., Barron, B., Engle, R. A., Erickson, F., Goldman, R., et al. (2010). Conducting video research in the learning sciences: Guidance on selection, analysis, technology, and ethics. *Journal of the Learning Sciences, 19*(1), 3–53. https://doi.org/10.1080/10508400903452884.

Engeström, Y. (2015). Learning by expanding. In *Learning by expanding* (2008) (2nd edn, p. 299). https://doi.org/10.1111/j.1440-1746.2006.04459.x

Gimmler, A. (this volume). The relevance of relevance for research ethics. In K. Otrel-Cass, M. Andrée, & M. Ryu (Eds.), *Examining research ethics in contemporary science education research*. New York: Springer Publishing Company.

Han, B.-C. (2018). *Hiperculturalidad*. Barcelona: Herder.

Johansen, G., & Anker, T. (this volume). Science education practices: Analysing values and knowledge when conducting educational research. In K. Otrel-Cass, M. Andrée, & M. Ryu (Eds.), *Examining research ethics in contemporary science education research*. New York: Springer.

Jones, A., McKim, A., & Reiss, M. J. (2010). *Ethics in the science and technology classroom*. Leiden: Brill\Sense. https://doi.org/10.1163/9789460910715.

Keshet, Y., Ben-Arye, E., & Schiff, E. (2013). The use of boundary objects to enhance interprofessional collaboration: Integrating complementary medicine in a hospital setting. *Sociology of Health and Illness, 35*(5), 666–681. https://doi.org/10.1111/j.1467-9566.2012.01520.x.

Latour, B. (2007). *Reassembling the social. An introduction to actor-network-theory*. Oxford: Oxford universtiy press.

Lovibond, S. (2002). *Ethical formation*. Cambridge: Harvard University Press.

Orlander, A. A., & Lundegård, I. (this volume). Sex education – Normativity and ethical considerations through three lenses. In K. Otrel-Cass, M. Andrée, & M. Ryu (Eds.), *Examining research ethics in contemporary science education research*. New York: Springer.

Reiss, M. J. (2008). Should science educators deal with the science/religion issue? *Studies in Science Education, 44*(2), 157–186. https://doi.org/10.1080/03057260802264214.

Ryu, M.-J. (this volume). Ethical considerations in ethnographies of science education: Toward humanizing science education research. In K. Otrel-Cass, M. Andrée, & M. Ryu (Eds.), *Examining research ethics in contemporary science education research*. New York: Springer.

Scoles, J. (2018). Researching 'messy objects': How can boundary objects strengthen the analytical pursuit of an actor-network theory study? *Studies in Continuing Education, 40*(3), 273–289. https://doi.org/10.1080/0158037X.2018.1456416.

Sjøberg, S. (2007). Science education: An interdisciplinary field. In K. Tobin & W. M. Roth (Eds.), *The culture of science education* (pp. 95–106). Rotterdam: Sense Publishers.

Star, S. L., & Griesemer, J. R. (1989). Institutional ecology, "translations" and boundary objects: Amateurs and professionals in Berkeley's Museum of Vertebrate Zoology, 1907–39. *Social Studies of Science, 19*(3), 387–420. Retrieved from http://www.jstor.org/stable/285080.

Tangen, R. (2014). Balancing ethics and quality in educational research – The ethical matrix method. *Scandinavian Journal of Educational Research, 58*(6), 678–694. https://doi.org/10.1080/00313831.2013.821089.

Jaume Ametller is a Serra Húnter Associate professor of science education at the University of Girona. He is interested on the design of teaching sequences and materials, on the role of communication in the construction of knowledge, and on how theory informs our understanding of how people learn.

Part II
Epistemological Considerations for Ethical Science Education Research

Chapter 8
Ethical Challenges of Symmetry in Participatory Science Education Research – Proposing a Heuristic for Ethical Reflection

Maria Andrée, Kerstin Danckwardt-Lillieström, and Jonna Wiblom

8.1 Introduction

The advancement of participatory methodologies and educational action research has raised challenges about research ethics that concern the relations between different actors. Different forms of participatory research rest on cooperation between teachers, researchers, and students in different forms of relations. The ways in which these relations are enacted are often related to research objectives, epistemology, who is involved, and the context in which the study is carried out (Wagner 2016). Sensevy et al. (2013) have proposed a symmetry principle as a device for guiding enquiry in teacher-researcher cooperation in mathematics education design-based research. According to this symmetry principle, all participants in a design based research project should share responsibility for the intervention, even if it is a teacher who carries out the teaching and thereby takes the minute-to-minute decisions in the intervention situation. Sensevy and his colleagues point to the value of a local, practical indistinguishability between the teacher and the researcher, where both the teacher and the researcher share responsibility for responding to the problem of teaching practice both in theoretical and concrete ways. Thus, participatory methodologies and educational action research involve ethical challenges beyond

M. Andrée (✉)
Stockholm University, Stockholm, Sweden
e-mail: Maria.andree@mnd.su.se

K. Danckwardt-Lillieström
Stockholm University, Stockholm, Sweden

Huddinge Municipality, Huddinge, Sweden

J. Wiblom
Stockholm University, Stockholm, Sweden

City of Stockholm, Stockholm, Sweden

© Springer Nature Switzerland AG 2020   123
K. Otrel-Cass et al. (eds.), *Examining Ethics in Contemporary Science Education Research*, Cultural Studies of Science Education 20,
https://doi.org/10.1007/978-3-030-50921-7_8

the questions of informed consent and confidentiality in conventional university-based research where educational practices are regarded as fields for data collection. In this chapter, we draw on the principle of symmetry to argue for a research ethics in participatory science education research based on the ontological, epistemological, and methodological value commitments of participatory research.

We seek to disentangle some ethical challenges emerging from three different teacher-researcher collaborations in science education research. What values are at stake and what are the potential tensions in attempting to secure different values? This includes the ethical implications of requiring shared responsibility between teachers and researchers in implementation and knowledge generation. We will provide three examples of how the principle of symmetry may be extended to function as a device for ethical reflection on value commitments at play in participatory science education research. The three examples are studies that we have been involved in ourselves and reflect different ways of and different struggles in enacting researcher-teacher relationships. The first example involves the attempts of a university-based researcher (Andrée) to establish research collaboration with an in-school teacher. The second example involves a researcher who is school-based (Wiblom) working to establish research collaboration with science teacher colleagues at her school. The third example involves a researcher pursuing research in her own classroom practice (Danckwardt-Lillieström).

8.1.1   Values at Play in Participatory Research

In research there are values at play that relate to ontology, epistemology, and methodology. The ontological commitments underpinning participatory action research include a democratic and egalitarian value base, a commitment to hold oneself responsible for how one tries to influence other people's learning, and acknowledging that one is part of the world connected to other people in an endeavor to undertake enquiry with others (McNiff 2017). In other words, participatory methodologies assume that the researcher is always part of the situation they are studying and that we as researchers need to negotiate our values and forms of living with others.

The epistemological values have to do with what counts as valuable in terms of knowledge and knowledge production (how truthfulness may be established). An epistemological value commitment in participatory action research is that knowledge is uncertain and ambiguous, and that knowledge about social situations is created through dialogue with one another (McNiff 2017).

Methodology refers to how the research is conducted. A strong methodological value commitment in participatory action research is that all practitioners, in our cases all the teachers participating in the research, are agents and not objects of study, recipients, or onlookers (Newton and Burgess 2008; cf. Carlgren 2012). In educational research, teachers are commonly viewed as 'practitioners' "trapped in a practical relationship to their work, while researchers hold a theoretical stance" (Sensevy et al. 2013 p.1032). This view has implications for how the process of knowledge production is understood – as a process involving a distanced researcher, positioned as the 'thinker', and a teacher, positioned as the 'doer' applying scientific results to practice. Participatory research seeks to overcome the classical dualism between 'persons who think' and 'persons who do', and instead contribute to affirming teachers as professionals and opening up new spaces for teachers to explore instead of bringing in outsider knowledge (McGlinn Manfra 2009; Price and Valli 2016). Or, as Bradbury-Huang (2010, p. 93) puts it, there is a striving to "take knowledge production beyond the gatekeeping of professional knowledge makers." Another methodological value commitment is the transformative orientation to knowledge creation (cf. Bradbury-Hwang 2010). The aim of the research is not just to seek understanding of a particular social situation but also to contribute to the improvement of it – in our case, science education classroom practices (cf. Elliot 1991; Carr and Kemmis 1986).

The ontological, epistemological, and methodological value commitments introduced above are foundational to the principle of symmetry. The principle assumes no practical or epistemic differences between different agents in the research. Sensevy and his colleagues (2013 p. 1033) write that "every agent plays 'her game', that is, proposes to the collective her first-hand point of view, what she 'sees' and what she 'knows' from her position, a point of view that is irreducible to any one other." Such an arrangement may foster what Sensevy and colleagues have termed a local practical indistinguishability where the involved agents take collective decisions and share ways of responding to a problem in a teaching practice. According to Sensevy and colleagues, the principle of symmetry is both epistemological and ethical. We would argue that the principle of symmetry also adheres to the ontological and methodological values of participatory research.

In the following we provide three examples of studies that we have been involved in ourselves with different configurations of teacher-researcher collaboration. In light of the principle of symmetry the three cases are examples of imperfection and struggle. In writing about the three cases we use first person plural (we) and singular (I) to voice our experiences as individual researchers and members of research teams. We hope that, by sharing our shortcomings and ethical reflections with a wider audience, we will contribute to making science education research more responsive to the tensions of values that are inevitably part of any research process.

8.2 Maria Andrée: A University-Based Researcher Attempts to Establish Research Collaboration with an In-School Science Teacher

The first case is a project initiated by university-based researchers. The project involved the first author and a research colleague at Stockholm University (Associate Professor Lotta Lager-Nyqvist) who set up an action research project in collaboration with a primary school teacher who taught science in the first and second grades.

The project was part of a larger externally-funded project on learning and narrative remembering where the research group worked with questions of how inquiry-based science education (IBSE) practices are, and potentially could become more responsive to students' experiences and funds of knowledge. Prior to the action research part of the project the researchers had published two analyses of opportunities for learning in IBSE teaching practices. The first study focused on how IBSE teaching practices constrained students' opportunities to draw on personal funds of knowledge (Andrée and Lager-Nyqvist 2012), and the second on how students engaged in informal spontaneous play in their work to transform the given tasks into something more personally meaningful to them (Andrée and Lager-Nyqvist 2013). With the new study the research team hoped to explore the conditions of classroom practice that previous studies had suggested were epistemically productive.

In the initial phase of the project, we contacted a local municipality to explore possibilities for developing a partnership with a couple of schools. After meeting with and presenting the ideas of our research group to the science teachers and the heads of two different schools, we decided to start working with one school. We were initially granted rapid access to a team of teachers and one of the teachers invited us to work more closely with her. She had substantial experience in teaching primary school although she was less experienced at teaching science since she had only recently completed a course on teaching science in primary school.

8.2.1 Challenges of Symmetry

Although access was granted rapidly, the process of establishing a partnership took much longer than we as university-based researchers had anticipated. Our starting point in the project was that we wanted to engage in participatory research and establish a non-hierarchical relationship with the collaborating teachers. We did not, as in much of our previous work, want to set the agenda and the design of the intervention beforehand. However, the participating teacher later admitted that for a long time she had been unsure about what we were after: What was the researchers' agenda? Over the first phase of 4 months we had several meetings with the teacher. However, our field work during this period became ethnographic work. After a while we started to get to know the school and the teachers and their worries and frustrations in relation to teaching science in primary school. At the beginning of the second semester we eventually started working as a team. We started to plan research lessons together, targeting issues such as how to create conditions for students to learn about inquiry in science education and how students' ability to talk about inquiry work could be developed as part of teaching. For the first time, as a group, at least to some extent, we came to share a research object and practical responsibility for the design and implementation. One of the inventions by the group was to use homework experiments to facilitate more explorative conversations about observation and interpretation among the students. In implementing the lessons, we divided the students into three groups and all three research team members took responsibility for enacting the intervention.

Although, as university-based researchers, we would subscribe to a democratic and egalitarian value base, we had no prior experience of establishing such a relationship. We first had to become part of the classroom situation, together with the classroom teacher and the students, before being able to engage in any kind of negotiation concerning classroom practice. In other words, we could not establish a symmetrical relationship with the classroom teacher before becoming part of the world of the classroom.

8.2.2   Divergent Objectives

In the process of knowledge production a divergence in objectives emerged between us as researchers and the teacher. The group produced a working report on the project. Although there was an attempt at challenging asymmetry in knowledge production, the university researchers ended up acting as the primary analysts and writers, thus reproducing asymmetry. In the working report, which was not formally published, the classroom teacher contributed with a preface on her experience of the project. Being unaccustomed to participatory research, as researchers we had some difficulty discerning the values of the project in terms of knowledge outcomes. We presented parts of the study at academic conferences, and years later a publication presenting a theoretically motivated interactional analysis of remembering as instructional work in science classroom practices was published as Andrée, et al. (2017). The scientific publication was primarily written to satisfy the needs of the larger, externally-funded project of which ours was a smaller part.

In the end, the action research project functioned primarily as a learning opportunity for us as university-based researchers in attempting to engage in participatory research. The starting point of the project was determined from a university-based researcher position which was reflected in the difficulties establishing symmetric relations in setting up, conducting, and reporting on the research project. In the beginning there was a clear lack of local practical indistinguishability, although this relation was eventually, at least partially overcome during the second semester when the group engaged in the collaborative planning and design of research lessons. Throughout the project tensions of value commitments included the lack of connectedness with the local classroom situation from the university-based researchers, challenges of epistemic ambiguity in the object of study, and difficulties achieving symmetry in the knowledge creation process. In hindsight, the team failed to reconcile the dualism of university-based researchers as 'thinkers' and the teacher as 'doer'.

8.3 Jonna Wiblom: A School-Based Researcher Establishing Research Collaboration with Teacher Colleagues in the School

The second case is part of my (Wiblom's) PhD project on the use of digital technology in upper secondary biology education. The PhD project was part of a graduate school on school-subject didactics in a collaboration between the local municipal and the university. The project was supervised by Maria Andrée (the first author) and Carl-Johan Rundgren at Stockholm University. The graduate school aimed to create opportunities for practicing teachers to systematically examine and explore didactic challenges while working part time as teachers in school. During the PhD project I functioned both as a school-based researcher (working on a PhD thesis while attending courses at the university) and as an upper secondary biology teacher in a public upper secondary school in Stockholm. My PhD project was conducted as a two-year design-based research study. During the phase of planning and implementation I worked with two biology teacher colleagues at the school. Of the three, I was the only one participating in a PhD program while my colleagues taught biology and science full time (during the second year they received some minor reduction of their teaching load for participating in project meetings).

During our first planning meeting we started by framing challenges in biology teaching that might be worth engaging in the project. The challenge we agreed upon had to do with our use of digital technology in our classrooms. The school could be described as an 'early adopter' regarding digital technology in education. All students and teachers had access to individual laptops and everyday classroom work across school subjects involved both teachers and students' use of digital technology. With digital tools (and not least the Internet) entering our classrooms, we needed to reinterpret and expand our understanding of the biology curriculum in relation to digitization. We asked ourselves questions about what knowledge of biology was useful for participation in contemporary societies. We were mutually frustrated with our use of digital technology as a means of facilitating canonical biology learning. For instance, we engaged our students in software-supported activities like taking lecture notes, writing lab reports, charting fieldwork data, or preparing oral presentations during class. The Internet was primarily used as a dictionary for the learning of biology concepts and as a resource for finding videos illustrating, for example, mitosis or photosynthesis.

We formulated the overarching aim of our research collaboration as an exploration of how to integrate digital technology with education so that it would facilitate students' learning of biology in ways that made it relevant to their participation in society. More specifically, the research objective was to design and implement classroom activities that developed students' capacity to critically and ethically examine science related information on the Internet, and further to produce science-related digital media themselves. Human physiology and health were chosen as curricular areas of content. The two interventions that followed were implemented as part of the regular biology teaching in the second year science classes of my colleagues (I was teaching first year students at the time). During the implementation my colleagues taught their classes and I was responsible for video documentation and for taking field notes. Throughout the implementation period we had weekly meetings to plan lessons, evaluate learning outcomes, and reflect upon student learning. After the implementation phase, I wrote and published the resulting research paper with my two supervisors without the involvement of my two teacher colleagues.

8.3.1 Challenges of Symmetry

Although we were all teachers in the same school and to some degree shared responsibility for the research there were different risks at play in terms of symmetry-asymmetry throughout the project.

During the two cycles of implementation, my colleagues and I were continuously confronted with contradictory demands that put our roles and work relationships at stake. My colleagues were responsible for the biology course, to ensure that the students were provided opportunities for reaching the prescribed learning objectives, and to assess their achievements. Their participation in the study put their trust and relationships with the students at stake as they were about to teach the rest of the biology course after the intervention had come to an end. It also turned out that the introduction of new digital activities challenged some parents' expectations and conceptions of school biology. At the end of the first intervention a parent contacted one of my colleagues and questioned how 'chatting' and 'producing websites' could possibly be part of students' learning science in school. My colleagues were also the ones confronted with students' questions and anxiety regarding examination. The following was part of a conversation that took place between one of my colleagues and two of her students as the new digital activities were introduced in biology class.

Teacher: You are going to discuss ethical aspects of health and evaluate health-related resources on the Internet.
Student: Is it like…in the social science program?
Teacher: Yes, but we will focus on issues that are extremely scientific.

In Swedish upper secondary biology curricula, students' ability to critically examine science-related issues in media from ethical standpoints is expressed as a central learning goal. The question as to whether they would be working "like… in the social science program," points to tensions between the introduced activities and the activities commonly associated with science education in school. My colleague answered that they would learn about "issues that are extremely scientific" emphasizing that the ethical discussions (and critical examination) would focus on the areas of human physiology and health. Her answer pointed to a view that the main value of engaging in critical and ethical examination of health issues in science education had to do with students' learning about human physiology. However, our research objective was formulated the other way around, namely to examine how students' engagement with health issues in media could support their development of subject specific capabilities in terms of critical and ethical examination of science. From the research perspective, I as a researcher was interested in expanding the notion of knowledge in science education. Rather than a strict focus on students' knowledge about "scientific issues", the research focus was on qualifying students' capabilities to engage critically and ethically with health-related issues in ways informed by science education practice.

Towards the end of the above lesson my colleague announced that the upcoming digital activities involved students' conversations about health-related ethical dilemmas over a chat forum, which would constitute grounds for assessment.

Student 2: Will you assess the websites?
Teacher: We will assess the chat [about health ethics dilemmas], *that* is the examination; a discussion on the Internet.
Student 2: Are you serious? A chat?
[...]
Student 3: But how should we chat then?

Introducing website production and chatting as part of biology class caused confusion and anxiety among the students. As illustrated in the transcript above, one student even questioned whether the teacher was serious, and another student asked, "but how should we chat then?" Then and there, we were puzzled by the fact that these sixteen-year-olds (presumed digital natives) would ask us as teachers how to communicate on social media. Looking back, however, the students' questions may be interpreted as expressions of uncertainty concerning what was to be expected from them as students of Biology when communicating online in the context of formal biology education.

Engaging the students in new digital activities such as website- and podcast production, required us as teachers and teacher-researcher, as well as the students, to reconsider and renegotiate existing epistemological value commitments regarding what counts as school biology. It also contributed to challenging the establishment of a symmetrical relationship in conducting the research since both students and parents made it clear that it was my colleagues who were to be held responsible for any risks associated with the implementation.

8.3.2 Divergent Objectives

Over time, divergent objectives regarding my and the teachers' engagement in the project evolved. The episode that follows is part of a conversation held at one of the research-team meetings in the first cycle of intervention during which forms for assessing students' achievements were discussed.

Teacher 1: Maybe we should have like one research objective and one objective regarding students' learning? There is no problem to relate to the curriculum regarding ethical perspectives and source critique. That renders great discussions, to become substantial and to practice those abilities at the same time. To be critical and to reason about ethical issues at the same time. Perfect! But relating that to their digital competences? That is problematic.

[...]

Teacher 1: I mean when we assess their capabilities to search for information, participate, and interact online, that doesn't really correspond to curricular learning goals. Their participation. That becomes very vague to assess. Exercise is good, but perhaps not assess. Interesting from a research perspective though.

The above utterances exemplify how tensions between different roles and interests in the research collaboration gave rise to uncertainty about how the project would contribute to improving the biology teaching practice. When my colleague distinguished between an interest in developing students' learning and a research interest, she pointed out that we had not succeeded in establishing and upholding a shared research objective. From my perspective, however, the student learning was not separate from the research objective, in fact, the student learning was the focus of the research. However, my colleague expressed a concern that renegotiating the biology curriculum learning goals in relation to digital activities would be problematic and not necessarily compatible with students' discussions about ethics and health. Based on the epistemological assumption that learning is situated and developed by participation in social practices, I advocated that students' achievements should be assessed *as* they engaged in the digital activities. As illustrated above, my colleague found such a perspective interesting; however, challenging to reconcile with established assessment practices (based on written assignment such as exams, reports, and short papers). To reconcile the tensions between the established teaching practice and the exploratory research objective, my colleague suggested that the students could be given the opportunity to "exercise" critical examination and ethical reasoning while engaging in digital activities. However, she also concluded that such an activity "becomes very vague to assess." In the end, the students' achievements were assessed by means of a written assignment where the students were asked to reflect on their own learning process.

In this case, the difference in risks involved for me as a researcher (although school-based) and the teachers, contributed to the emergence of an asymmetric relationship. We did share local responsibility in planning and engaging with the students during the lessons during the implementation, but we did not share the consequences in terms of accountability for assessment and in relation to students and parents. We also did not share accountability for outcomes in terms of publication.

8.4 Kerstin Danckwardt-Lillieström: A School-Based Teacher Researcher Conducting Interventions in Her Own Practice

The third case is an example of where I (Danckwardt-Lillieström), the researcher, was not only school-based but also the teacher responsible for conducting the interventions in my own practice. I have extensive experience in teaching science in upper secondary school (over 20 years) and participated in a science education graduate school at Stockholm university with combined financing from the local municipality and the Swedish Research Council. The purpose of the graduate school was to strengthen the link between research and school development and to expand science teachers' didactic knowledge base and capacity for didactic analysis of school practice. The core activities of the graduate school program included developing and testing models for teaching science in school (Wickman et al. 2018).

This study focused on teaching chemistry; the choice of subject grew out of experiences in upper secondary school of chemistry being a subject that students often find challenging, abstract, and hard to grasp. In collaboration with my two supervisors (the first author Maria Andrée and Margareta Enghag), I investigated ways in which creative drama could be used to support students' chemistry learning in upper secondary school. The study was framed as design-based research focusing on how creative drama could be used to afford students' learning of abstract, non-spontaneous chemical concepts. Theoretically, the study was framed within social semiotics, exploring what kind of semiotic work the students were engaged in in different enactments of creative drama in chemistry education; for example, how the students used their own bodies as semiotic resources to make sense of the concepts of chemical bonding and electronegativity. The study was conducted in three cycles during 2015 and 2017. The first and second cycle were enacted by me acting as a teacher and researcher in my own practice. The third cycle was later implemented in collaboration with a chemistry teacher in another school. In this chapter I will focus on the challenges of symmetry as both researcher and teacher in my own classroom.

8.4.1 Challenges of Symmetry

The first two cycles of the design-based study of creative drama were conducted as part of my ongoing chemistry teaching. During these cycles I was both teacher and researcher and, thus, related to the students both in their roles as participants in daily classroom life and as participants in research. In this context there is no asymmetry in terms of a hierarchical relationship between researcher and teacher or dualism between the thinker and doer in the research. However, new ethical challenges are invoked in terms of the values at stake in conducting the research and I had to deal with these challenges in both my roles. To conduct a study as a researcher where my own students were participating required careful reflection on the responsibility and values at stake. The challenges of symmetry concern the balancing of interests relating to me as a researcher and me as a teacher and risks pertaining to the entanglement of these interests.

One fundamental ethical challenge has to do with student-teacher dependency. In a situation where a teacher is also a researcher there will be dependency related to assessment and future grading. In the following I reflect upon an episode in the classroom a week before the intervention of the second cycle. Prior to the start of the chemistry lesson, the students had been informed that the next lesson would be part of a research project aimed at contributing to the development of learning in chemistry. The students were given information about the project and written consent forms were handed out to the students. The students were invited to read the information and decide whether to provide written consent to be filmed during the research. As I was walking around the classroom, I overheard one student asking another if she was going to participate in the research and the student answered that she wanted to help me with the research. The student thus expressed that she is giving her consent to participate in the research in order to help me. In other words, the consent becomes an expression of support for me as a teacher. Would the student have given her consent to an independent researcher with whom she had not had a relationship? Does the student agree because she is dependent on her teacher and fears that the assessment of her skills could be adversely affected if she declines to participate? Or, does the student agree because she is sympathetic towards her teacher? According to the Swedish national regulations on consent, the participants in a research study have the right to decide for themselves about their participation, and in an investigation with active contributions from the participants, consent shall always be obtained (Swedish Research Council 2017). As a researcher you have an obligation to ensure that participation in research is voluntary. One challenge here is to find ways to circumvent dependency. This will have implications for when, how, and possibly by whom students are informed about participation in a research study. To avoid undue influence, the researcher would commonly present a study to the students and not their teacher, but in this case I was both.

8.4.2 Divergent Objectives

To me as a teacher and researcher there have been different objectives at play. As a chemistry teacher, my primary objectives have to do with student learning and the development of my local school practice. As a researcher, my primary objective has to do with the production and sharing of new knowledge about chemistry teaching. In hindsight, the research project has influenced both the actual teaching and the ways of talking and thinking about teaching in the local school. As a teacher, I have been able to take the newfound insights generated through research directly into teaching practice and to share these insights with colleagues who do not participate in research but still indirectly transform and develop their teaching as an outgrowth of the research process. There is thus an added local transformative value to conducting research as a teacher (cf. Newton and Burgess 2008).

This design-based study was conducted as part of the ongoing teaching. Hence, transformation involved both students and teachers. Careful consideration has been given to those students who were in the classroom but who, for different reasons, did not want to participate in the study. Regarding this issue of participation in the research, there are clearly divergent objectives. In relation to the objective of teaching chemistry, I as a teacher was accountable for the learning of all students and great caution was given to designing the intervention so that the students not participating in the research were guaranteed an equivalent instruction. In contrast, the objective of my research project focussed on the participation of the students participating in the research intervention.

As a teacher I also have an ethical-moral obligation to provide students with equal learning opportunities. Thus, it becomes important that the students who do not give consent are provided equal learning opportunities, particularly when interventions are conducted as part of the ordinary chemistry teaching. Throughout the three cycles of the study, the students who did not consent to participate in the research study were placed at a table beyond the reach of the audio and video recorders. Thus, these students could participate in all learning activities. The rationale for this was that the teaching activities were designed to the best of our knowledge and the activities were expected to contribute to the understanding of the students. It is of the utmost importance that the teacher, from an ethical perspective, ensures that the students who chose not to participate in the study are treated equally. I emphasized that the students should not to be left out and I made sure to address this group of students during whole-class interaction and, as the teacher, to listen to their discussions and provide them with opportunities to ask questions. Thus, I strove to be attentive to their participation in the classroom conversation. The problem of the participation of students who do not give consent presents in all interventionist research; however, the difference for me as both teacher and researcher was that I had overall responsibility for the learning of all students, whereas an independent researcher would be accountable primarily to the participants of the study.

Another issue of concern relating to divergent objectives, given my dual roles as both researcher and teacher, had to do with how I dealt with sensitive information during the process of analysis of the collected data. Through the process of data analysis, a researcher is given a unique opportunity to peek into the social micro-dynamics of the classroom. As a teacher you normally do not gain access to private student conversations. For a teacher-researcher, listening to these recordings not only provides insights related to the object of study – in this case conceptual learning in chemistry – but also into the social micro-dynamics of the classroom, including tendencies towards abusive behaviors and bullying. In this study, I was made aware of several examples of such demeaning social micro-dynamics. In one group, the analysis of video data displayed how a girl, who was often very silent in the classroom, participated in other nonverbal ways in the group work. The film also revealed that she had very low status in the group. On one occasion she clearly signaled that she wanted to visualize an atom with her body, but the group ignored her and chose another group member to participate in the dramatization. In another group, one student uttered racist comments about another student on several occasions. Being able to gain insight into the behavior of group members towards each other can enable the teacher to consciously change the group composition and pay attention to the silent students in a way that promotes learning. This may also be part of the development of the educational design.

In comparison with a researcher who is temporarily in the classroom, who does not really know the students, and is not able to follow up on the research with the students, the teacher-researcher finds herself in an ethically delicate situation. The teacher-researcher is bound to the students' consent to participate in the study, which in this case had to do with the use of drama in chemistry education and its potential for conceptual science learning. The students had also been informed that the data collected would be used for research purposes only. On the other hand, the teacher-researcher is the adult teacher in the classroom responsible for ensuring that the classroom environment is free from abusive behavior in all forms. If abusive behavior had caught my attention during the activity it would have been common practice to act on it as a teacher. The situation becomes somewhat different when the covert abusive behavior is first recognized weeks or months after the teaching activity, when I, as a researcher, listen to and transcribe the recordings. One way to resolve the tension between values would be to say that the teacher-researcher should not begin analyzing data before the end of the school year. On the other hand, this would be a way of concealing the tension regarding the responsibilities one has as a researcher not to go beyond the given consent and the responsibilities one has as a teacher to care for the students.

In relation to the objectives of research, a common objection to conducting research in my own teaching practice concerns my own bias – whether the results may have been influenced by my own opinions, backgrounds, and values. This might be interpreted as an issue of divergent objectives, involving tensions between goals of producing knowledge and goals of developing educational practice. For example, Barab and Squire (2004) emphasizes that the role of the researcher in the design experiment may endanger the validity. If the researcher is personally involved in the design, development, and implementation it is challenging for the researcher to ensure the accuracy of the results. On the contrary however, we maintain that the researcher's participation has been beneficial in this study. Anderson and Shattuck (2012) and Coe (2013) recognise that the participation of the researcher presents a well-known challenge for many forms of qualitative research where it is difficult to ignore the bias of the researcher in the research process. However, Anderson and Shattuck (2012) argue that the researchers themselves, with their bias, insight, and deep understanding of the context, are the best research tools. In this design-based study, I have been not only the teacher and researcher in the same person, but I have also conducted the study in my own classroom with my own students, as a teacher-researcher. Therefore, I have a deep knowledge of the context in which the interventions are conducted, which may be a great advantage in terms of validity in the analysis. For example, the importance of background knowledge in observing students made it possible for me to compare how the students participated in chemistry teaching during the intervention compared with how they participated in the traditional teaching context. Thus, when it comes to the question of bias and deep understanding of the context, I argue that this does not really have to do with divergent objectives but is rather a question of epistemology.

8.5 Conclusions

These cases of teacher-researcher collaboration illustrate how participatory methodologies involve a transformation of ethics and epistemology and how the ethics and epistemology become intertwined. In the different forms of research, the start and the end of research, as well as the aims and roles of teachers and researchers are blurred in different ways. The scrutiny of these three examples of research collaboration reveals the limitation of traditional ethical considerations that focus on the integrity and consent of the individual – like those provided in guidelines from national authorities (such as Swedish Research Council 2017). In addition to the standard ethical reflection on informed consent and confidentiality, an ethics of participatory research in science education has to include considerations of the ontological, epistemological, and methodological values at stake.

8.5.1 Ontological Values

The ontological values include a commitment to hold oneself responsible for how one tries to influence other people's learning and acknowledging that one is part of the world connected with other people in an endeavor to undertake enquiries with others (McNiff 2017). Thus, acknowledging and reflecting upon the transformative potential of participatory research becomes a necessity. There will always be issues of hierarchies and dependencies between participants in an interventionist research project even if the principle of symmetry is a highly-held value. This includes dependencies and hierarchies between university-based researchers and teachers, between researching teachers and other teacher colleagues, and between researching teachers and students.

For the different actors there are different issues at stake in participatory research that will need negotiation regardless of the configuration of the actors. For example, assessment will be at stake in an intervention for students and teachers. For any interventionist study in a natural setting the question of assessment requires ethical reflection. Teachers have obligations in relation to the assessment and grading of students, whereas researchers will not necessarily, for example, prioritise questions of assessment in the design of an intervention targeting student learning. For science teachers, the science content may also be at stake if the research intends to challenge established practice. This was very much the case in Wiblom's PhD project. For a teacher-researcher participating in a PhD program, one issue will be about maintaining working relationships with colleagues and students in the school beyond the PhD program and, not least, making sure sufficient data is produced to write up a thesis. For a university-based researcher there are similar issues at stake – for example, issues relating to expectations of publication.

8.5.2 Epistemological Values

The epistemological values assume that knowledge is uncertain and ambiguous; created through dialogue with one another (McNiff 2017). In particular, the question of validity of the results of a specific study is a question about what counts as data and what data will be sufficient for making an epistemological claim.

The second case illustrates how participatory research may involve epistemological tensions including researchers, teachers, students, and parents. Using participatory research methodologies to transition from a Vision I of science education in terms of mastering a set of concepts and educating professional scientists to a Vision II of science education as a citizenship education may be challenging in terms of design of teaching and assessment practices (Roberts and Bybee 2014). Epistemological tensions revealed through participatory research may be difficult but of fundamental importance to negotiate in that the tensions may correspond to objectives of transforming education practice by means of research. In the second case, the research object had to do with developing students' capacities for critical scrutiny and ethical examination of science in media through science education. The research was premised by an objective of transforming educational practice in line with a conceptualisation of science education within Vision II. Thus, it would have been valuable to have stated the epistemological assumption of what may count as biology in order to open up for dialogue with the participating teachers, students and parents. Unless some shared sense of epistemology is negotiated it is difficult to imagine how symmetry and a shared object of research might be established at all.

8.5.3 Methodological Values

A transformative orientation to knowledge creation is foundational to participatory research (cf. Bradbury-Hwang 2010). This includes a commitment to transforming, improving local science teaching practices, and to transforming and expanding the professional knowledge base of the teaching profession. Traditionally, important methodological considerations have included informed consent, confidentiality, and protection of participant integrity. Participatory research becomes even more delicate.

The research is embedded in social practice with social relations of dependency and loyalty between participants and different issues are at stake for participants in different positions. When it comes to the question of informed consent, reflection is required on the conditions under which consent is provided: what are the power-relations at play and how can such power relations be managed to minimize damage to the participating students and teachers? When it comes to the question of confidentiality and integrity there are delicate issues of how to deal with richness of data and, for example, what to do with information about abusive behavior.

Finally, participatory research also requires ethical reflection on anonymity versus credit. If research interventions are designed and conducted in teams of university-based researchers and school-based teachers, it becomes problematic from an ethical point of view if only the university-based researcher is credited in the publications that stem from the intervention.

8.5.4 A Heuristic for Ethical Reflection on Participatory Research

Based on the principle of symmetry as proposed by Sensevy et al. (2013) and our experiences of conducting participatory research in science education, we propose a heuristic for ethical reflection on participatory science education research expanding the principle of symmetry in the dimensions of ontology, epistemology, and methodology (see Fig. 8.1). To function as a device for ethical reflection on the value commitments at play in participatory science education research, we have formulated reflective questions in relation to the dimensions of ontology, espistemology and methodology. The ontological values demanding reflection includes holding oneself responsible for how one tries to influence other people's learning in relation to acknowledging the embeddedness of research in social practice as well as acknowledging the values at stake for participants in different positions. This ontological reflection will be as relevant for a university-based researcher collaborating with an experienced teacher as for a teacher taking on both the roles of researcher and teacher in her own classroom practice. The epistemological values demanding reflection concerns what is required to make epistemological claims. Central aspects to the epistemological reflection concerns the negotiation of a shared (valid) research object as well as shared objectives or visions for science education. In this chapter we have provided examples of how such negotiations might be challenging. The methodological values demanding reflection concerns the social relations of dependency between researchers, teachers, students, parents and others, questions of what to use as data (and what not to use), how data should be analysed and interpreted but also how to reconcile potentially contradictory values of a symmetric relationship in relation to ownership of the outcomes of the research and requirements from conventional ethical standards of integrity and anonymity.

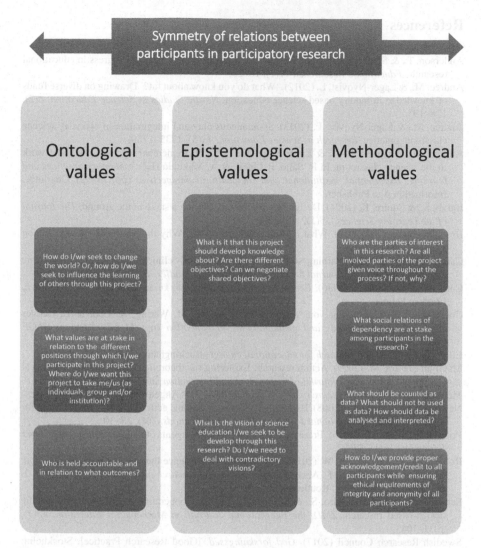

Fig. 8.1 A heuristic for ethical reflection in participatory science education research

8.5.5 *Concluding Remark*

Engaging in ethical reflection on our work has provided us with opportunities for becoming more responsible about how we conduct research. We hope that sharing our shortcomings and reflections with a wider audience will contribute to making science education research more responsive to the tensions of values that will inevitably be part of any research process.

References

Anderson, T., & Shattuck, J. (2012). Design-based research. A decade of Progress in educational research? *Educational Researcher, 41*(1), 16–25.

Andrée, M., & Lager-Nyqvist, L. (2012). 'What do you know about fat?' Drawing on diverse funds of knowledge in inquiry based science education. *Nordic Studies in Science Education, 8*(2), 178–193.

Andrée, M., & Lager-Nyqvist, L. (2013). Spontaneous play and imagination in everyday science classroom practice. *Research in Science Education, 43*(5), 1735–1750.

Andrée, M., Wickman, P.-O., & Lager-Nyqvist, L. (2017). Remembering as instructional work in the science classroom. In R. Säljö, P. Linell, & Å. Mäkitalo (Eds.), *Memory practices and learning: Experiential, institutional, and sociocultural perspectives* (pp. 75–92). Charlotee: Information Age Publisher.

Barab, S., & Squire, K. (2004). Design-based research: Putting a stake in the ground. *The Journal of the Learning Sciences, 13*(1), 1–14.

Bradbury-Hwang, H. (2010). What is good action research?: Why the resurgent interest? *Action Research, 8*(1), 93–109.

Carlgren, I. (2012). The learning study as an approach for "clinical" subject matter didactic research. *International Journal for Lesson and Learning Studies, 1*(2), 126–139.

Carr, W., & Kemmis, S. (1986). *Becoming critical: Education, knowledge and action research.* London: Falmer.

Coe, R. (2013). Conducting your research. In J. Arthur, M. Waring, R. Coe, & L. Hedges (Eds.), *Research methods and methodologies in education* (pp. 41–52). London: SAGE Publications Ltd.

Elliot, J. (1991). *Action research for educational change.* Buckingham: Open University Press.

McGlinn Manfra, M. (2009). Action research: Exploring the theoretical divide between practical and critical approaches. *Journal of Curriculum and Instruction, 3*(1), 32–46.

McNiff, J. (2017). *Action research - all you need to know.* Los Angeles: Sage.

Newton, P., & Burgess, D. (2008). Exploring types of educational action research: Implications for research validity. *International Journal of Qualitative Methods, 7*(4), 18–30.

Price, J. N., & Valli, L. (2016). Preservice teachers becoming agents of change. *Journal of Teacher Education, 56*(1), 57–72.

Roberts, D. A., & Bybee, R. W. (2014). Scientific literacy, science literacy and science education. In N. G. Lederman & S. K. Abell (Eds.), *Handbook of research on science education* (Vol. 2, pp. 545–558). New York: Routledge.

Sensevy, G., Forets, D., Quilo, S., & Morales, G. (2013). Cooperative engineering as a specific design-based research. *ZMD, the International Journal on Mathematics Education, 45*(7), 1031–1043.

Swedish Research Council (2017). *God forskningssed.* [Good Research Practice]. Stockholm: Vetenskapsrådet.

Wickman, P.-O., Hamza, K., & Lundegård, I. (2018). Didaktik och didaktiska modeller för undervisning i naturvetenskapliga ämnen. *NorDiNa, 14*(3), 239–249.

Wagner, J. (2016). The unavoidable intervention of educational research: A framework for reconsidering researcher-practitioner cooperation. *Educational Researcher, 26*(7), 13–22.

Maria Andrée is an Associate Professor of Science Education at the Department of Mathematics and Science Education at Stockholm University, Sweden. Drawing primarily on socio-cultural theory, her research focuses on science education practices and the conditions for students' participation and learning, particularly in relation to questions of science curriculum, scientific literacy and citizenship. She primarily works with ethnographic and design-based research studies in science education. She has also pursued a line of research concerning the involvement of external actors in science, technology and mathematics education. She is currently one of the scientific leaders of *Stockholm Teaching & Learning Studies* – a platform for research in collaboration with teachers – designed to initiate, support and conduct small-scale classroom-based didactic research on teaching and learning. She has published in peer-reviewed journals including *International Journal of Science Education, Research in Science Education, Cultural Studies of Science Education*, and *Journal of Curriculum Studies* among others.

Kerstin Danckwardt-Lillieström is a PhD-student and teacher of chemistry and biology in upper secondary school in Huddinge municipality, Sweden. She also works as a research coordinator in a project aimed at developing participatory educational research in the local municipality. She has participated in a research school focusing didactic modelling in science education at the Department of Mathematics and Science Education, Stockholm University. She conducts design-based research on the use of drama in upper-secondary chemistry education. Her latest publication was titled "Creative drama in chemistry education. A social semiotic approach" which was published in *Nordic Studies in Science Education*.

Jonna Wiblom is a PhD student and lecturer at the Department of Mathematics and Science Education, Stockholm University. Her research focuses the development of students' capabilities to navigate and critically examine science related issues in contemporary media. She has published her research in peer-review journals, including *Science & Education* and *Reserach In Science Education*. Her background is in the teaching of biology and science studies in upper secondary school. Alongside her PhD project she has been part of *Stockholm Teaching and Learning Studies*, a platform designed to support teachers' participation in didactic research.

leaders of Stockholm Teaching & Learning Studies – a platform for research in collaboration with teachers – designed to initiate, support and conduct small-scale classroom-based didactic research on teaching and learning. She has published in peer-reviewed journals, including *International Journal of Science Education*, *Research in Science Education*, *Cultural Studies of Science Education*, and *Journal of Curriculum Studies*, among others.

Karim Hamza-Lilliestråle is a PhD student and teacher of chemistry and biology, in upper secondary school in Huddinge municipality, Sweden. She also works as research coordina- for an upper-level project at developing participatory. She has also didactic research in the local municipality. She has participated in a research school focusing didactic modelling in science education at the Department of Mathematics and Science Education, Stockholm University. She conducts design-based research on the use of drama in upper-secondary chemistry education. Her latest publication was titled "Creative drama in chemistry education: A social semiotic approach", which was published in *Nordic Studies in Science Education*.

Jonna Wiblom is a PhD student and lecturer at the Department of Mathematics and Science Education, Stockholm University. Her research focuses the development of students' capabilities to navigate and critically examine science-related issues in contemporary media. She has published her research in peer-review journals, including *Science & Education*, and *Research in Science Education*. Her background is in the teaching of biology and science studies in upper secondary school. Alongside her PhD project she has been part of Stockholm Teaching and Learning Studies – a platform designed to support teachers' participation in didactical research.

Chapter 9
Living Authenticity in Science Education Research

Jennifer D. Adams and Christina Siry

9.1 Introduction

In 1989 Egon Guba and Yvonne Lincoln established a set of criteria on which to judge the authenticity and ethics of qualitative research, the Authenticity Criteria (AC). These criteria were developed in response to the positivistic assumptions of internal and external validity, reliability, and generalizability that guide quantitative research and often extend to the judgement of qualitative research. The AC are responsive to research paradigms that recognize subjectivity and the context-dependent structures that mediate research outcomes. Such research requires a hermeneutic/dialogic approach that places the researcher in the context and requires her to be aware and reflective of how stakeholders experience and interpret their lived experiences in relation to the research context. With this chapter, we intend to underscore the necessity of paying careful attention to the AC, in order to increase the possibilities that all stakeholders can learn, grow, and benefit from engaging in the research process.

We are science education researchers grounded in cultural studies, and we adopt dialogic, participatory approaches in an effort to try to centralize participant perspectives. As such, the AC are critical to our work. The sections that follow elaborate how paying attention to catalytic, educative, ontological and tactical authenticity (Guba and Lincoln 1989) in the research process can facilitate transformations in educational contexts, including classrooms and institutions. The process of research reflexivity extends to the teaching and learning process, as researchers become

J. D. Adams (✉)
University of Calgary, Alberta, Canada
e-mail: jennifer.adams1@ucalgary.ca

C. Siry
University of Luxembourg, Esch-sur-Alzette, Luxembourg

© Springer Nature Switzerland AG 2020   143
K. Otrel-Cass et al. (eds.), *Examining Ethics in Contemporary Science Education Research*, Cultural Studies of Science Education 20,
https://doi.org/10.1007/978-3-030-50921-7_9

mindful that all participants ought to benefit from the research, as well as from everyday science teaching and learning, both in formal as well as informal settings.

This chapter emerges from conversations at a scholarly writing workshop at the University of Luxembourg, in which we sought to define a research agenda to foster innovation and invite collaboration in science education research. Forty participants from 12 different countries came together for a multi-day workshop, in which participants interacted in small groups on particular foci relevant to the cultural studies of science education. We (authors Christina and Jennifer) were in a group charged with examining issues of equity and social justice in science education research by using our respective research lenses and understandings. Prior to our meeting, we each wrote short reflections on our current research interests and shared them with each other. This enabled us to begin the time together in our workshop by discussing the individual reflections and identifying cross-cutting themes and challenges to doing ethical, equity-oriented research. Our enthusiastic conversations used words that evoked a dialogic, passionate, stakeholder-focused, social justice-oriented research approaches. Yet, when we later created a collective Wordle of our individual five-page descriptions of research, we were surprised to find that it presented quite a different picture. This picture was one of traditional education research, with words such as "knowledge," "schools," and "issues," being the most prominent words that had emerged from our writings. Given that Wordles are created based on word frequency, we were surprised by the resulting image, which stood in contradiction to how we envisioned our own work. This discrepancy led us then to question why our research, as written on paper, was so different from how we presented our research in conversation? This prompted us to revisit Guba and Lincoln's (1989) AC, which we claim grounds the equity-focused work we do, and then ask a larger, more critical, question of our work: Are we living the Authenticity Criteria in how we practice science education research? This question is all the more important if we are thinking about how descriptions of equity and corresponding equitable practices need to be reconsidered, and how we can work to meet the needs of diverse learners, which was one of the key questions we aimed to address in the workshop. Furthermore, we are in a paradigm that seems to place greater value on research that has an experimental design rather than on the kinds of naturalistic work that we do. As such we felt that it was even more important to ensure that our research was aligned with criteria that allows us to measure the "truthfulness" and the impact of our research, and show us where we need to work harder with this alignment. Furthermore, we wanted to ensure that our research is ethically sound and includes both consent and beneficence of all stakeholders involved.

Emerging from our reflexive analysis of this contradiction, in this chapter we draw on our experiences to reflect on what the AC provides for science education research. Further, we consider how the AC is one component of authentic inquiry and discuss the ways that we have extended these criteria towards more ethical practices in our research and research relationships. We use the following guiding questions to frame our dialogue:

- Starting with the AC, what does it mean to do "stakeholder-focused" research?
- What is missing from the AC that may deepen our engagement in research and strengthen the work that we do?
- How can we apply the AC to the communication/dissemination of our work, in order to highlight the necessity of working towards research that centers and benefits participants?

As we address these questions, our theoretical elaborations weave text written in the genre of metalogue throughout the chapter. Metalogue was developed by Gregory Bateson as a type of dialogic text to allow respondents to explore individual and collective epistemologies about a topic (Maran et al. 2011; Roth et al. 1998). Herein, we use this genre to highlight our individual research foci, while engaging in a written dialogue to further our theoretical and methodological understandings on the notion of ethical practices in science education research. We do so with a particular lens on how the AC can serve as a path towards critically grounded, ethical work with participants. The AC, which we elaborate more in the following section, mediate the ethical dimensions of the research that we each do, as they facilitate research focused on the stakeholders themselves, as we highlight from exemplars in our own research throughout the chapter.

Our use of the term "stakeholders" highlights that we aim to empower research participants through the research process. This is in contrast to positivistic research approaches that extract data from the context with the assumption of the neutral stance of the researcher and generalizability of research. With stakeholder-focused research we recognize that research is contextual, and that students, teachers, parents and administrators all can benefit (or be harmed) by the research process. It is our role as researchers to ensure that benefits are realized, harm is minimized, and that, in turn, research contexts move towards more equitable practices. The AC criteria can serve as a heuristic that affords this stance, as we elaborate in the sections that follow.

9.2 The Authenticity Criteria: Our Roadmap for Stakeholder Centered Research

The AC were developed by Guba and Lincoln in response to the positivistic assumptions of internal and external validity, reliability, and generalizability that guide quantitative research and which were often extended to the judgment of qualitative, narrative research. Lincoln (1995) describe authenticity criteria as,

> highly reflective of the commitment of inquiry to fairness (balance of stakeholder views), to the learning of respondents as much as to the learning of the researcher, to the open and democratic sharing of knowledge rather than the concentration of inquiry knowledge in the hands of a privileged elite, and to the fostering, stimulation, and enabling of social action. (p. 277)

The AC were conceived as a heuristic for researchers to engage in stakeholder-centred research in naturalistic settings. These criteria are responsive to research paradigms that recognize human subjectivity and context-dependent structures that mediate research outcomes. The following are brief definitions of the key tenets of the AC (Guba and Lincoln 1989):

(a) **Fairness**, the extent to which the understandings of all stakeholders are accounted for in the research. This is determined by an assessment of the extent to which all competing constructions have been accessed, exposed, and taken into account in the research process, that is, in the negotiated emergent construction.

(b) **Ontological** authenticity, the extent to which the knowledge of individual stakeholders is informed and changed as a result of participating in the research. This is determined by an assessment of the extent to which individual constructions (including those of the evaluator) have become more informed and sophisticated.

(c) **Educative** authenticity, the extent to which individual stakeholders gain an understanding of the perspectives of others. This is determined by an assessment of the extent to which individuals (including the evaluator) have become more understanding (even if not more tolerant) of the constructions of others.

(d) **Catalytic** authenticity, the extent to which the research facilitates changes in the behavior of the stakeholders; stakeholders are empowered towards agency and transformation in relation to the research. This is determined by an assessment of the extent to which action (clarifying the focus at issue, moving to eliminate or ameliorate problems, sharpening values) is stimulated and facilitated by the evaluation.

(e) **Tactical authenticity,** the extent to which stakeholders are empowered to take action that the research implies or proposes. This is determined by the actual actions that stakeholders take towards change.

Jenn: My research focuses on equity in science teaching and learning through studying identity, relationships to places/contexts, informal science education and creativity. My current project is learning about teacher identities in relation to informal science learning. As a researcher, the AC offered me a framework to think about my research beyond just learning about teacher identities in relation to informal science education, but also to view how teachers transform meanings of informal science education to match their identities, goals and teaching contexts. Furthermore, it has been critical for me to extend the AC to encompass authentic inquiry, which Konstantinos Alexakos (2015) describes as research that is holistic, recursive and emphasizes multiple viewpoints and voices, and resists the theory/practice and research/findings dualisms. This has allowed my research to be responsive to changes that happens as stakeholders learn and expand their individual and collective agency as a result of participating in the research. For example, the cultural construct of race in general and Blackness in particular became important in understanding both how their teaching identities emerged during their first years of teaching and how informal science was redefined and enacted both for the collective and vis-a-vis the unique contexts of their individual schools.

Chris: Being responsive to changes that happen in your research is an important contribution that the AC bring to your work Jenn, and the AC can provide a support for better understanding the subjective experiences of the participants. In my own work, my research team and I examine classroom interactions to learn about how children explore, interpret, and discuss their world. In doing so, we are learning much about children's imaginative, creative, and complex notions of science phenomena. This stands in contrast with notions of science as a primarily fact-based subject, a perspective often represented in curricula as well as being common in teaching practices at the primary level. The contrast between teaching practices that reproduce a factual perspective of science, with a focus on answers, and children's engagement in science as a way of investigating their world, with a focus on questions, is used to problematize a discrepancy between how science is taught and how it is engaged in by children. This discrepancy is particularly troubling in the multilingual classrooms of Luxembourg, a country with an immigrant population of approximately 50% (MENJE 2014), and over 150 different nationalities represented. Over the past 25 years, our country has experienced "the highest sustained inflow of immigrants with respect to the total population" in Europe (Eurydice 2004). As this diversity is also represented in schools, many classrooms have children that speak a multitude of languages, but not often the languages of instruction. In unpacking this discrepancy, my research team and I seek to examine how equitable practices might need to be reconsidered to meet the need of diverse (young) learners, and the AC provide a structure for considering this.

The AC recognizes that "education research with human subjects must benefit those who are involved in the study and that researchers have a responsibility to those who agree to be involved that benefits will not be realized only in the future, but will also lead to improvements as the research is enacted" (Tobin 2015). While these criteria are not meant to be formulaic, they provide a heuristic for researchers who endeavour to do stakeholder-centered and justice-oriented research. In such research contexts, the learning is ongoing and benefits are realized at multiple levels by multiple stakeholders. These criteria also foster research approaches that are collaborative and actively engage stakeholders in the knowledge production process so that they are empowered by participating in the research.

Jenn: Similar to you Chris, the AC is a structure for centering equity in my work. I received a National Science Foundation grant to do the teacher identity research. Because I had to articulate a "hypothesis" and narrow framework of how I was defining identity, I went into the project with limited view of the role of how classrooms, schools and students shape teacher identity. I was focused on the practices that they learned in informal settings and how it would be enacted in the classroom. However, because my research approach used methodologies that emphasize stakeholder agency (i.e cogenerative dialogues) the AC came more to the forefront of how I conducted the research. With this, the view of myself, my research team and my teacher participants were expanded to examine both the nuances of identity and how equitable practices emerged for my teachers because of an increased reflexive examination of who they are as social beings in relation to their students. In addition, I emphasize critical frameworks that highlight the power relationships that structure social life. This, combined with dialogue and AC allowed all of us, researchers and teachers, to become more aware of institutional structures that were barriers to equitable science learning and sought ways, both individually and collectively, to resist, challenge and transform those structures.

Chris: Combining critical theory with authentic research approaches as you mention is a salient way to advance equity in our work. Using analyses grounded in hermeneutics and critical theoretical perspectives, the research that my team and I conduct seeks to reveal approaches to working together with teachers and children towards finding openings, in which the spaces created by the apparent conflict between teachers' expectations and children's realities actually becomes a space of productivity and possible transformation. We work to create structures for teachers to come together reflexively, as they discuss their own science teaching and their students' interactions around science learning opportunities. Prolonged engagement of the researchers and a continual process of participatory engagement of the teachers (through shared data analysis, for example), are central to ensuring that the AC are being met throughout our research process. Teachers are integral to this, as they are encouraged to participate in planning the direction that the research takes, and their perspectives and ideas are central to how our research unfolds over time.

9.3   Critically Grounded Authentic Educational Research: Understanding Why

Our work is guided by critical theoretical perspectives that have been forwarded by Joe Kincheloe, Shirley Steinberg and Paulo Freire amongst others, and grounded in cultural studies approaches as emphasized by William Sewell and Kenneth Tobin. As such, we work to recognize and learn from the contextual complexities of doing research in science education, and we draw on methodologies that can support working with participants in research projects that seek to facilitate transformation. Research can be positioned as the production of knowledge; in science education this is the production of knowledge around science teaching, learning and engagement as well as questioning the foundations of science as taught in schools and out-of-school institutions. Towards that end, Jürgen Habermas (2015) proposed three knowledge-producing purposes, analytical, hermeneutic and critical, each with a different central foci and outcomes. Paul Terry (1997) has linked these purposes to the types of questions that can be approached in research, including:

- An analytic interest in knowledge production supports empirical outcomes ("knowing that")
- A hermeneutic interest in knowledge production supports understanding ("knowing how") and
- A critical interest in knowledge production supports emancipation ("knowing why").

Further, he suggests that these three types of knowledge production purposes can provide a key to understanding education structures (Terry 1997). We believe that it is not enough to know that something has happened, or even to know how it happened. To conduct research grounded through critical perspectives requires us to work towards understanding why it has happened. Only once we understand the why, can we work systematically towards changing practices to be more socially just. We thus conceive that such points on the outcomes (knowing -that, -how, and -why) can also serve as guide for considering the purposes of educational research. With an eye on considering the purposes of research projects with participants, we ask ourselves what is the goal of doing research and what interests are being supported through particular lenses and approaches? With our work situated through critical perspectives we seek to work towards "the question of transcending the existent" (Young 1992, p. 31), which means that we work to unpack meanings and evidences of learning that are taken for granted in order to allow for an expanded view of what is valued as knowing and knowledge production. This is even more critical in science education where science is often positioned as objective, with often deeply embedded cultural notions about who can legitimately participate in the scientific endeavor.

Chris: Recently I came across an article that underscored, for me, the core of what it means to work towards authentic research praxis, as the authors wrote that "Authenticity involves an assessment of the meaningfulness and usefulness of interactive inquiry processes and social change that results from these processes" (Shannon and Hambacher 2014, p. 1). It is precisely the assessment of the meaningfulness as well as the social change that is for me the goal of utilizing the AC to reflect upon the research process. I believe that research should support change and transformation, and it is a goal that I hold dear, as I work to create spaces for teachers to "see" the immense capacities young children have for engaging in science. Young children ask questions about many things that adults may take for granted (Opdal 2001), and as such, my research has sought to highlight children's "wonderings" (e.g., Siry 2013) to both illustrate the diversity and complexity of children's questions, and also to deconstruct the questions together with teachers and teacher education students in order to support recognizing the value of young children's questions and ideas. When we layer onto these different ways of engaging with science the complexities of multilingual classrooms, there is often a tension that emerges between teachers' expectations for science teaching and learning, and children's engagement in science. This contradiction is one that I have increasingly noticed as I work in classrooms, and the AC (Guba and Lincoln 1989) have provided a foundation to examine how to work within, across, and around these tensions so that the spaces between the teachers and the students becomes productive, and ideally transformative.

Jenn: Meaningfulness, usefulness and social change are important goals to have for conducting research. In my research both with secondary teachers and science faculty, stakeholder-focused research means that while I am learning from and about their identities as teachers or practices as faculty, I am also creating the context that allows them to learn more about their own teaching and desires for transformative learning experiences for their students (cf. Adams 2007). In both instances, I do research and facilitate professional learning. While I learn about them I am also creating a space for them to learn more about themselves and their professional development needs by connecting them with resources to grow and expand their practices towards creativity and equity.

Chris, you also mentioned recognizing and working within the contradictions, this is critical in equity work. In my current work with faculty there is the tension between creativity and assessment. The former tending towards expansive practices and engagement in science learning while the latter towards the rote memorization of discrete facts. The latter is what is expected of students as they advance to upper-level courses – the emphasis on knowing existing scientific knowledge rather than engagement in scientific knowledge production. So, as a faculty community of practice we have ongoing discussions on how to work these tensions while advancing an agenda of creativity. Stakeholder-focused research allows me to center the voices and lived experiences of the faculty and the unfolding learning that occurs in the community; experiencing research as a lived event rather than as a fixed project with a definitive start and end.

9.4   From Authenticity Criteria to Authentic Inquiry: Collaborating with Participants

Our interest in applying the AC in our research stems from participating in Kenneth Tobin's research squad at the Graduate Center, CUNY. There we learned not only how to ask the critical questions of "what is happening" and "why is it happening" as a means of eliciting thick descriptions (Geertz 2008) of our research contexts, but also as a way of engaging stakeholders as a praxis of including their voices in these descriptions. For example, many of us use cogenerative dialogues, "a form of structured discourse in which [stakeholders] engage in a collaborative effort to help identify and implement positive changes in [a given teaching and learning context]" (Martin 2006, p. 694) as a research methodology and praxis to afford stakeholder agency and transformation around the research topic. We have engaged in co-writing and co-researching with participants, and in doing so have explored the necessity of seeking different perspectives in the research process (e.g., Siry and Zawatski 2011), the value of co-teaching for professional development (e.g., Siry and Lara 2012) and ethical implications of collaborative research (Siry et al. 2011). With equity and social justice as central to our work, it is important to create and learn from contexts that allowed stakeholder to engage in meaningful and relevant science learning experiences. Furthermore, we value critical approaches that recognize the political nature of knowledge and knowledge production along with the power dynamics that exists in institutional structures that often serve to maintain societal status quo, for example structures that contribute to urban schools having inadequate resources for rich and expanded science learning. Using the AC as a heuristic for engaging in critical, social justice-oriented research affords opportunities for all stakeholders to participate in learning about and improving science teaching and learning.

Jenn: When I and my colleague Preeti Gupta enacted research on youth identity in relation to working in a science center as Explainers, (see Adams and Gupta 2013), we emphasized the learning aspect of the project. It was both about learning about youth science-related identity and the young people learning both from each other and the researchers about being better informal science educators. Keeping stakeholders' voices central to the research allows us to both gain a deeper understanding of various sociocultural issues in relation to science teaching and learning as well as affording participants agency in transforming these contexts to be more meaningful, relevant and socially just. In my research about science teacher identity, race became a very important social construct that shaped teacher identity and enactment and the teachers worked both individually and collectively to create learning contexts that afforded success in science for their Afro-Diasporic and Latinx students of color (who in the United States remain underrepresented and underperforming in science). Centering diverse voices and perspectives provides a deeper understanding of contexts for teaching and learning and how different people participate in learning. The AC lends itself to research approaches that are collaborative and multi-perspectival. The AC supports authentic inquiry that centers the well-being of all stakeholders in research. Central to authentic inquiry is agency and transformation; it is expected that the research contexts and stakeholders should change from participating in the research; if the research is done correctly all stakeholders should gain a deeper understanding of the educational context and issues at hand and collectively work towards improving teaching and learning for all. Tobin (2015) notes, "authentic inquiry addresses additional values concerning ethics and acknowledges that all knowledge is inherently political, reflecting participants' in social space" (p.12).

Chris: Those examples illustrate how the AC can afford guiding constructs for conducting research that is fair, transformative, and equitable. As researchers in two diverse international contexts, careful attention to the AC allows us to make science education research a more participatory process and increase the possibilities that all stakeholders will learn, grow and benefit from engaging in the research process. By focusing on ontological, educative, catalytic, and tactical authenticity in the research process, classrooms, contexts and institutions have the potential for transformation. This process of research reflexivity extends to the teaching and learning process, as researchers become mindful that all participants are benefitting from not only research, but from everyday science teaching and learning, both in formal and informal settings.

Keeping the AC central to our work allows us to be reflexive researchers who aim to continuously reflect on our commitment to improving science teaching and learning. With this reflexivity we strive towards fairness, in that we seek to facilitate research in which all participants' voices are heard and considered. As we move through the research process with our participants, we intend to continually and recursively assess the authenticity of the work we are doing. For us it is important to consider participants' awareness of the complexities of their social environments (ontological authenticity) as well as the extent to which they express increased awareness and respect for the perspectives of others (educative authenticity). These can be facilitated through methodologies that work towards dialogic encounters amongst all stakeholders (Shannon and Hambacher 2014). Catalytic and tactical authenticity both require focusing on change as there ought to be empowerment that results from the research, however, this can be difficult to assess as it is not always easy to "see" such changes. Catalytic authenticity is evident if there was action stimulated by the stakeholders, and tactical authenticity implies a redistribution of power (Shannon and Hambacher 2014). The reflexive space that the AC affords allows us to create dialogic structures that welcome participant voice, and that mediate participants' agency and the potential for action on the part of stakeholders.

Chris: As a science education researcher and educator the AC provide me a lens to ensure that the research I engage in with young children and their teachers is as fair and equitable as possible. With the methodologies I have adopted in my work, I seek to be as participatory as possible, and I strive to work towards gaining a multiplicity of perspectives on science education in the primary school classes that my team and I conduct research with. Strategies that we implement to support the AC include positioning participants as central to the research and working together to collect a diversity of data resources that serve as points for individual and collective reflection and discussion. Keeping an eye on the AC throughout the research process underscores the necessity of providing participants a voice, and most importantly, considering and reacting to participants input. In doing so, we work together to create structures that mediate teachers' and children's agency in the teaching and learning of science.

Jenn: Similarly, the AC provide me with a heuristic to ensure the rigor of my research and allow for a framework that emphasizes collective learning and stakeholder agency. With my research team we emphasize dialogues in our data collection methods. We had two groups, teachers who participated in 45-min interviews about their practice and a longitudinal group of (then) new teachers with the goal of learning how their teaching identities and corresponding practices unfolded during their first years in the classroom. For the first group, we developed interview questions that prompted an exchange between the researcher and teacher participants rather than a didactic question and answer approach. This allowed for a natural conversation about teaching to unfold and both the researcher to exchange ideas and examples about science teaching and learning. In the ongoing group we used a cogenerative dialogue approach (Martin 2006, see above) and this allowed for authentic voice and creating a community of learners rather than the strict dichotomy of researcher and researched. As the lead researcher and with the AC in mind, I have an ongoing concern of the learning of stakeholders – that they are learning from participating in the research as myself and my research team are learning about the focus of the research. In the teacher research, this means that they are continually learning to teach, in the case of this research, learning how to adapt and use available resources to meet their teaching and learning goals. Also affording them the space to develop their own meanings about what it means to teach science in a diverse, urban context.

9.5 Relationality, Trust and Well-Being as Emerging Authenticity Criteria

Earlier we posed the question of "what is missing from the AC may deepen our engagement in research and strengthen the work that we do?". This question allowed us to first consider that the AC is in many ways a living document. It was created nearly three decades ago and while it is both seminal and relevant, as we continue to transform research to be more authentic to and in teaching and learning spaces, these criteria will also evolve. In other words, if we conduct research that is true to the AC, it is expected that both research and subsequently the AC will change in order to both mirror and validate new research/educational contexts. Here we reflect on or research and discuss related emergent criteria.

Jenn: Chris, we talked about trust and relationality and how it is important to first build trust between the researcher(s) and research participants and amongst research participants. Trust is described as, "people's willingness to be vulnerable due to their confidence that the individual(s) they interact with are open, benevolent, reliable and honest" (Id-Deen and Woodson 2016, p. 45). Trust is developed through positive interactions and ownership with the foundations of trust based on shared expectations, persistence, commitment, and voice (Ennis and McCauley 2002). Extending this to research, we could see where a criterion of trust (beyond trustworthiness) would be important in creating and maintaining a research context that authentically affords the equal participation of all stakeholders. This would attenuate the power dynamics that often exists between the researcher and researched and teacher and students if all are working towards a sense of safety and mutual respect for all stakeholders.

Relationality also resonates with my research. With relationality described as, "relational dynamics shape processes of partnering and the possible forms of learning that emerge in and through them" (Bang and Vossoughi 2016, p. 174), I can see where relationality, combined with trust has enabled much more expanded learning both for the research participants and for my research team. For example, creating a space of mutual trust has allowed the teachers to develop a sense of agency in how they define and enact informal science learning in their classrooms. I enacted what Rita Kohli (2014) refers to as "reciprocal vulnerability" where I shared my own experiences with learning to teach science in the same urban contexts, including my successes and failures, in order to create a space where the teachers felt safe and comfortable in sharing their own. Listening to their experiences allowed me to expand my definition of informal science learning and think more deeply about teaching enactment, in terms of the transformation of resources at hand to envision new learning opportunities for students, and its relationship to identity. This has also been important in my work around creativity – creating a safe and trusting space to allow faculty to discuss their practices, feel agency in how they are conceptualizing creativity and begin to push the boundaries of their thinking about what is possible in the classroom. I do this within a community of practice framework and with the AC in mind I establish it as a cogenerative space where, as the facilitator/researcher, I emphasize shared meanings and encourage risk-taking through trying and sharing new pedagogical enactments in their classrooms and labs.

Chris: I am in complete agreement with you Jenn, as I believe that trust and relationality are the cornerstones of ethical research. Neither of these are ensured by the AC however, and we should take a moment to reflect on this. The AC emerged from a time in education research that was predominantly guided by positivistic paradigms, and there was a need to develop criteria to "match" criteria from such positivist paradigms, such as validity and reliability. As such the AC help us work towards research that is authentic and hopefully transformative. What is missing however is an explicit focus on the well-being of the participants in that they feel they are in a safe environment to express their ideas and have them be heard as well as respected. Trust and positive relations are at the heart of ensuring well-being, and thus I personally feel that while the AC help us get there, we have to always be vigilant that we are taking care of those who are research participants.

Jenn: Chris, you articulated it well with "explicit focus on well-being," this critical not only for our research participants but also for our research teams. As leaders of research teams, we have to create spaces of trust and safety that allow our teams to also grow and expand as researchers. I believe that we have learned as much as we have about equity in our respective projects because we value and center perspectives our both our co-researchers and research students. The different lenses that they bring to our research process allow for us to develop expanded views of constructs like identity, agency, language interactions and learning. The AC have been a guiding heuristic but trust and relationality have extended the scope of work and lens of equity that we collectively apply to our research.

9.6   Extending the AC to Dissemination Research Practices: Highlighting the Value

Central to the AC is the learning and agency of all stakeholders. This does not end with the actual research but continues with communication and dissemination practices. This means that not only should others outside of the immediate stakeholders benefit from the research but also that all people have equitable access to the products of dissemination.

Jenn: This is an ongoing issue for me because so much of what we do in the academy is counted by the number of publications in "high impact" journals. However, if we think of our stakeholders, in my case classroom teachers, how will they access this research? It is not that teachers do not read educational research, but unless they are in graduate school, their reading is relegated to practical applications to the classroom; knowledge that will help them to be more effective at teaching. Furthermore, for science teachers, it is becoming increasingly important for them to keep up with advances in the subject area. Based on my experience of being a science teacher and later a science teacher educator both in a museum and in a university setting, I have learned that teachers learn by doing and through dialogues with other educators. Teachers learn to teach when they have opportunities to engage in the same activities that they are expected to do with students while thinking about adaptations to their unique learning contexts (i.e. Adams and Branco 2016). The teachers in my research have been learning from other research participants and from sharing their work at local and national teacher education conferences. These are key points of dissemination where teachers are able to share their understandings and enactments of informal science education with other teachers towards wider equity in science teaching. The educative tenet of the AC emphasizes the importance of educating others, beyond the immediate research participants, of what is learned in the research. So, I think we need to think about the key stakeholder audiences and their common channels of communicating and accessing those in order to reach broader members of given stakeholder audiences. For teachers this means engaging in professional development and targeting practitioner-oriented media.

Chris: The implications from my research point to the complexities of the way science is framed as a content area in school curricula and the ways in which science is 'done' by children in the primary school years (e.g., Siry and Lara 2012; Siry 2018). A contribution from this work is to underscore the need for supporting teachers and teacher education students in rethinking science education as a discipline in order to provide more equitable, authentic, practices for reaching diverse learners. However, as mentioned, the publications that typically "count" in the academy tend not to be the ones that practitioners might be reading. The AC compel us to find other venues for disseminating our work so that there is an educative component beyond the research-oriented publications. As such, it is important to find spaces that value teachers' and students' perspectives; presentations, publications, exhibits, demonstrations, etc. and encourage the research participants to actively engage in collaborative disseminations. I have written previously about the value of co-writing with students and research participants (e.g., Siry and Zawatski 2011) and my experiences have underscored the necessity of doing so. As we put thoughts to word and then to paper together, we share a creation of new meaning, and this is a powerful tool towards transformation in my experiences, one that is guided by the AC.

Jenn: It also behooves us, as stakeholders, and due to our status as mentors and leaders; gatekeepers in the academy, to take a critical stance on what counts as research. It is important for us to create opportunities for our students and mentees to disseminate research in ways that are valuable to stakeholders and to allow that to "count" as productivity in the academy. We would not be holding ourselves accountable to the AC if we maintain the status quo of focusing our communications within the academic community.

9.7 Dialogues and Reflections Towards Action: Closing Thoughts

We began this chapter by introducing the AC as a central focus in our research processes, and we have sought to add context and meaning through the use of the genre of metalogue. For us, the AC are a foundation for conducting authentic inquiry (Tobin 2015). Authentic inquiry as a methodology is grounded on the AC and relates to the well-being of all stakeholders in research. Using the AC as a guide, authentic inquiry has agency and transformation as central outcomes for research; as such, it is expected that the research contexts and stakeholders should change from participating in the research. If the research is focused on equity, then diverse stakeholders should gain a deeper understanding of the educational context and issues at hand and collectively work towards improving teaching and learning for all.

Chris: Stakeholder-focused research, as we have described above, requires reflexivity throughout the research process. Jenn, you said that teachers in your research come together to reflexively consider connections between informal science learning opportunities and their own identities, which is one critical component of working together towards equitable practices. In my work, we seek to create dialogic spaces with a goal of working towards a sort of reconceptualization as well, focused on what it means to teach science at the primary school level, and as in your work, this process begins first with creating a space that is open and responsive to participants' voice and reflection.

Jenn: I agree, dialogues and reflection are critical, and creating the spaces that allow for this to happen is paramount. This has been important not only in my research with teachers but also in working with my research team. I believe that this is similar to your team, Chris in that you create a space for your researchers to build their own theoretical lens while contributing to the overall knowledge production of the team. In my teacher identity work, my team engaged in a collaborative diffractive analysis where we avoided identifying themes and patterns that would bind the data but rather engaged in a "constant, continuous process of making and unmaking…arranging, organizing and fitting together" (Jackson and Mazzei 2012, p.1) where we applied different theoretical frameworks to the same data set. So, this open space of equity is central to my way of engaging with my research team.

I also feel that the AC, with its emphasis on stakeholder agency, has become embodied in my way of doing research, so for new projects I always seek to center stakeholders voice and perspective. For example, on a current research project, which focuses on creativity and science teaching and learning in postsecondary settings, the AC has allowed me to create a stakeholder-centered space where science faculty contribute to learning about creativity in science education while enacting it in their teaching spaces. I present existing frameworks that we discuss and they consider how it translates into their practices. This has allowed a community of practice with a shared vision of developing and enacting creative practices in science education to grow.

Our work with teachers, faculty and students has underscored for us the necessity of using the AC as a start towards assessing if we are truly working towards ethical and transformative research. But much as with the Wordle we generated in 2014, speaking and writing about these issues is not enough, as otherwise the picture that is represented of the research process is one that can reproduce positivist paradigms. Rather, the AC should serve as a continual reminder to ensure that the research process benefits all stakeholders and that research is authentic to the multitude of experiences of participants and that it mediates change as well as transformation towards more equitable science education practices.

References

Adams, J. (2007). The historical context of science and education at the American Museum of Natural History. *Cultural Studies of Science Education, 2*(2), 393–440.
Adams, J. D., & Branco, B. (2016). Extending parks into the classroom through Informal Learning and Place-based education. In P. Patrick (Ed.), *Preparing informal science educators*. The Netherlands/Finland: Springer.
Adams, J. D., & Gupta, P. (2013). "I learn more here than I do in school. Honestly, I wouldn't lie about that": Creating a space for agency and identity around science. *The International Journal of Critical Pedagogy, 4*(2).
Alexakos, K. (2015). *Being a teacher| researcher: A primer on doing authentic inquiry research on teaching and learning*. Dordrecht: Springer.
Bang, M., & Vossoughi, S. (2016). Participatory design research and educational justice: Studying learning and relations within social change making. *Cognition and Instruction, 34*(3), 173–193. https://doi.org/10.1080/07370008.2016.1181879.
Ennis, C. D., & McCauley, M. T. (2002). Creating urban classroom communities worthy of trust. *Journal of Curriculum Studies, 34*, 149–172.
Eurydice. (2004). *Integrating immigrant children into schools in Europe*. Brussels: Eurydice Network.
Geertz, C. (2008). Thick description: Toward an interpretive theory of culture. In *The cultural geography reader* (pp. 41–51). London: Routledge.
Guba, E. G., & Lincoln, Y. S. (1989). *Fourth generation evaluation*. London: Sage publications.
Habermas, J. (2015). *Knowledge and human interests*. New York: Wiley.
Id-Deen, L., & Woodson, A. N. (2016). "I know I can do harder work": Students' perspectives on teacher distrust in an urban mathematics classroom. *Urban Education Research & Policy Annuals, 4*(2).

Jackson, A. Y., & Mazzei, L. A. (2012). *Thinking with theory in qualitative research*. Hoboken: Taylor & Francis.

Kohli, R. (2014). Unpacking internalized racism: Teachers of color striving for racially just classrooms. *Race Ethnicity and Education, 17*(3), 367–387.

Lincoln, Y. S. (1995). Emerging criteria for quality in qualitative and interpretive research. *Qualitative Inquiry, 1*(3), 275–289.

Maran, T., Martinelli, D., & Turovski, A. (Eds.). (2011). Introduction. In *Readings in zoosemiotics* (Vol. 8). Berlin: Walter de Gruyter.

Martin, S. (2006). Where practice and theory intersect in the chemistry classroom: Using cogenerative dialogue to identify the critical point in science education. *Cultural Studies of Science Education, 1*(4), 693–720.

Ministère de l'Education national, de l'Enfance et de la Jeunesse (MENJE). (2014). *Les chiffres clés de l'Éducation nationale : statistiques et indicateurs – Année scolaire 2012–2013*.

Opdal, P. M. (2001). Curiosity, wonder and education seen as perspective development. *Studies in Philosophy and Education, 20*, 331–344.

Roth, W. M., McRobbie, C. J., & Lucas, K. B. (1998). Four dialogues and metalogues about the nature of science. *Research in Science Education, 28*(1), 107–118.

Shannon, P., & Hambacher, E. (2014). Authenticity in constructivist inquiry: Assessing an elusive construct. *The Qualitative Report, 19*(52), 1–13.

Siry, C. (2013). Exploring the complexities of children's inquiries in science: Knowledge production through participatory practices. *Research in Science Education, 43*, 2407–2430. https://doi.org/10.1007/s11165-013-9364-z.

Siry, C. (2018). The science curriculum at the elementary level: What are the basics, and are we teaching them?. *Thirteen questions for science education*. Peter Lang.

Siry, C., & Lara, J. (2012). "I didn't know water could be so messy": Coteaching in elementary teacher education and the production of identity for a new teacher of science. *Cultural Studies of Science Education, 7*, 1–30.

Siry, C., & Zawatski, E. (2011). "Working with" as a methodological stance: Collaborating with students in teaching, writing, and research. Invited contribution to special issue of. *The International Journal of Qualitative Studies in Education, 24*(3), 343–361.

Siry, C., Ali-Khan, C., & Zuss, M. (2011). Cultures in the making: An examination of the ethical and methodological implications of collaborative research [26 paragraphs]. *Forum Qualitative Sozialforschung/Forum: Qualitative Social Research, 12*(2), Art. 24. http://nbn-resolving.de/urn:nbn:de:0114-fqs1102245.

Terry, P. R. (1997). Habermas and education: Knowledge, communication, discourse. *Curriculum Studies, 5*(3), 269–279. https://doi.org/10.1080/14681369700200019.

Tobin, K. (2015). The sociocultural turn in science education and its transformative potential. In C. Milne, K. Tobin, & D. DeGennaro (Eds.), *Sociocultural studies and implications for science education* (pp. 3–31). Dordrecht: Springer.

Young, R. (1992). *Critical theory and classroom talk* (Vol. 2). Multilingual Matters.

Jennifer D. Adams is a Tier 2 Canada Research Chair of Creativity and STEM and Associate Professor at The University of Calgary where she holds a dual appointment in the Department of Chemistry and Werklund School of Education. Her research portfolio includes teacher identity, youth learning and identity in informal science contexts, with a focus on underrepresented youth and place/identity in transnational communities and environmental education. Her work emphasizes critical and sociocultural frameworks and participatory, qualitative, poststructural approaches.

Christina Siry is Professor of Learning and Instruction at the University of Luxembourg. Her research examines science learning and teaching at the primary and pre-primary levels, and she focuses in particular on the ways in which young children make and express meanings in interaction. Her work has been published in a variety of international journals, and she is currently a coeditor of the journal *Cultural Studies of Science Education*.

Chapter 10
Beyond Dichotomies/Binaries: Twenty-First Century Post Humanities Ethics for Science Education Using a Baradian Perspective

Kathryn Scantlebury ⓘ and Catherine Milne ⓘ

10.1 Introduction

To do an analysis in the post human is to embrace an ethical responsibility of attending to vibrant matter in all aspects of our projects and mapping the forces (or actants) as an agentic assemblage (Jackson and Mazzei 2016, p. 105).

Heretofore science education researchers have rarely used post humanistic theory to frame their research and its outcomes. For a discipline that focuses on the teaching and learning of science, that is, the natural world consisting of matter, science education researchers have consistently ignored how matter and humans are entangled in knowledge production and the ethics associated with those practices. In a 2012 interview, Karen Barad described ethics as an "accountability for the lively relationalities of becoming, of which we are a part" (p. 69) rather than the form of ethics with which most education researchers are more familiar, "right responses to a radically exteriorized other". With this set of quotes Barad captured both existing modernist understandings of ethics, which separate the researcher from the researched as the "exteriorized other", and post-human ethics, which make a claim for accountability based on human entanglement with the world and in doing so she has redefined the meaning of accountability. Fundamental to Baradian theory is the assumption that matter has agency, entities are not independent or privileged but come into being when they intra-act with each other, establishing agential realism (Barad 2007). Being ethical in this context means being responsive to the "pattern and murmurings" of matter with which we are entangled (Barad 2014, p. 3). Barad's theory challenges current norms and thought in science education research, and in

K. Scantlebury (✉)
Department of Chemistry and Biochemistry, University of Delaware, Newark, DE, USA
e-mail: kscantle@udel.edu

C. Milne
New York University, New York, NY, USA

© Springer Nature Switzerland AG 2020 159
K. Otrel-Cass et al. (eds.), *Examining Ethics in Contemporary Science Education Research*, Cultural Studies of Science Education 20,
https://doi.org/10.1007/978-3-030-50921-7_10

this chapter we seek to address the influence of material and matter on research into the teaching and learning of science. Using material feminism, we explore the ethical implications for science education researchers when the human and non-human are entangled, and we take seriously "who matters and what counts" (Taylor 2017, p. 5). As we noted, Barad's stance towards ethics is atypical to how ethics is often evoked in research that includes science education research and to how many institutions, which oversee and approve research studies, have instantiated ethical practices. Indeed, the idea of troubling the existence of universal ethical rules is a relatively recent development.

10.2   Modernity and Ethics

In the United States, most institutions take ethical stances that emerged from modernity, a contemporary ideology but one with a long history. Modernity put humans first and relied on the idea that human activity was socially constructed and that human action working through modern social institutions could transform society producing a grand narrative of progress from a point of privilege (Giddens 1990). The expansion of Western culture was used to justify and impose the human values and knowledge forms that are the basis of modernity. Stories of progress presented modernization as the path to advance knowledge, establish human rights, support democracy, create wealth and control nature for the benefit of humanity. Modernism is truly a humanist perspective with the notion of humanity, an abstract unifier of differentiated humanness, placing humans above all other living things and definitely above non-living matter (Rouse 1991). In modernity, the disciplines of science and technology are viewed as truly modern and therefore less open to critique than disciplines that emerged before modernity, which helps to explain why many scholars, including Donna Haraway and Karen Barad who trouble this belief, provide some relief from such rhetoric.

  In order to communicate a sense of why we find Baradian ethical perspectives so powerful and compelling, we first want to explore ethical perspectives that populate the institutionally approved requirements associated with research. Within modernity, a number of different ethical perspectives have been endorsed starting with teleological ethics from teleos, Greek for final purpose (May 1980). For example, you assume you are engaged in research if the end of the process in which you are involved is the discovery of new knowledge. For teleological ethics, this process of creating new knowledge also represents the fundamental good of a research life and does not have to be justified based on its contribution to any other good, the good is intrinsic not instrumental. Operating within this ethical framework would lead practitioners to an expectation that society should support research that generates new knowledge without any expectation that it would contribute to society in instrumental ways. Many science researchers, (for example, Marie Curie (Milne 2011)), positioned themselves in this way and continued to do so in the arguments they make for funding support for research into what are considered by the discipline fundamental ideas or areas of science. From a teleological perspective, the purpose of such research is inherently ethical so its methods also do not need justification. As William May (1980) notes teleologically, deception would not be considered an appropriate method if one's purpose was to generate new knowledge. Thus, according to teleological ethics, the strategies one uses to generate truthful knowledge must also be virtuous and appropriate for generating such knowledge. From a teleological perspective research is inherently ethical and valued because its purpose is to present the truth. As May notes, a teleological ethicist would be critical of the growth in medical ethics when compared with the dearth of people exploring moral issues related to race and disproportionality. Often, even with the best will in the world, economic imperatives bias the emphasis on specific areas of research. However, teleological ethics is both humanist and individual ensuring any understanding of the world generated from such an ethical approach is both limited and privileged.

A variation of teleological ethics but one applied to the collective rather than the individual is consequentialist ethics and the form of consequentialist ethics often applied to research conducted by institutions such as universities is utilitarianism. With its collective vision, this form of ethics argues for the greatest good for the greatest number. As an abstracted collective, the notion of 'humanity' aligns with the collective vision of utilitarian ethics. Within utilitarianism the focus is on the ethical character of the research methods used rather than the purposes of the research. Thus utilitarianism contributes to constructs of research ethics, such as *informed consent* and *confidentiality*. Although the connections to teleological ethics may be difficult to identify initially, consider that with respect to utilitarianism, the knowledge produced as a result of research is considered still to be intrinsically good or at least value neutral, so there is an expectation that the methods used to generate data will also be ethical (Milne 2011). Within utilitarianism, scholars can be expected to compare the risks to those being researched against the benefits of the research as they are asked to attest that the research does not present risks beyond those of everyday life.

In the United States, utilitarian ethics became formalized as ethical principles incorporated into research involving "human subjects" through the Belmont Report released by the National Commission for the Protection of Human Subjects of Biomedical and Behavioral Research (1978). Ethical methods in the Belmont Report included 'justice', which was defined in terms of benefit and burden with respect to the impact of research on participants. In the letter to the US President, the Chairman of the Commission, Kenneth Ryan, noted, "Publication and dissemination of this policy will provide Federal employees, members of Institutional Review Boards (Federally mandated ethics committees tasked with reviewing and approving applications to conduct research studies), and scientific investigators with common points of reference for the analysis of ethical issues in human experimentation" (National Commission for the Protection of Human Subjects of Biomedical and Behavioral Research 1978, no p.n.). Accepting the human bias of utilitarianism, a limitation of such ethics is the unequal distribution of power in the research relationship. Often the interventionists (as May calls them) are presented as benefactors to a recipient research population but while it is clear that the goal of the benefactors is that they benefit from the research, it is not clear that the recipients also benefit. May (1980) notes that "utilitarianism seems to defy what common sense associates with ethics: principles that have a categorical force irrespective of the results" (p. 363).

Within modernity we also have deontological (from the Greek for duty) ethics, an ethics that, rather than focusing on outcomes or process, as utilitarianism does, argues for universal ethical principles (Kant 1785/1938). The principles, which form the core of deontological ethics, apply equally to everybody, including researchers, and across time and space (they are categorical). Deontological ethics are truly human-centric because the duty is located in the structure of human reason. A limitation of deontological ethics is there is no room for nuance or context. Rather, these ethics are applied without fear or favor and as such create logical conundrums which highlight the limitations of such ethical stances.

10.3   Rejecting Grand Narratives: Postmodernism and Ethics

The emergence of postmodernism and the rejection of all grand narratives resulted in a critical scrutiny of these three forms of modernity ethics, teleological, deontological and utilitarian (Lyotard 1984). Jean-Françios Lyotard (1984) argued that rather than being a source of human freedom, modernity had supported the emergence of a modern West that was imperialistic and used its capacity with science and technology to subjugate other people making them into its image, a process we might call colonization. Of course, the narrative of progress that is integral to modernity implies a homogeneity that hides internal inequities applied to categories such as gender, race, class and sexual orientation. These approaches to ethical practices are also ones that a Baradian theorist would not endorse because while all disciplines should be committed to "helping make a more just world" (Barad 2012, p.153) this can only happen by being materially immersed and inseparable from the material world, being in and of. *Being in* means that we acknowledge our entanglement with research and its practices in an ethico-onto-epistemological context. The very construing of new knowledge involves us in ethical approaches as we are entangled with the living and the non-living. But such entanglement is not captured by modernist perspectives which elevate and separate humans from all other forms of matter, living and non-living.

10.4   Post-Humanism Ethics

Barad (2003) laid down the challenge for the material turn with her introductory sentence "Language has been granted too much power" (p. 801). Language is associated with the social aspects of culture and typically matter becomes enslaved to language. Barad noted the 'folly' of ignoring matter and challenged representalism because it was associated with acceptance of dualisms that separate humanity from the material world as we represent the real world in our minds assuming that these representations are reflections of reality. She proposed that agential realism could be used to understand how "the nature of the relationship between discursive practices and material phenomena, an accounting of "nonhuman" as well as "human" forms of agency", supplement an understanding of the precise causal nature of productive practices that takes account of the fullness of matter's implication in its ongoing historicity." (p. 810). Other researchers, such as Estrid Sørensen (2009), argue that acknowledging the epistemological and ontological significance of matter requires researchers to completely rethink their research and the ethics associated with that change. She argues, "Only by forgetting about human aspiration is it possible to start dealing seriously with materiality. Instead of beginning with the question of whether technology does what humans want it to do, we should ask how materials participate in practice and what is thereby performed" (p. 28). And one strategy for doing this is to begin with agential realism, "an epistemological-ontological-ethical framework that provides an understanding of the role of human and nonhuman, material and discursive, and natural and cultural factors in scientific and other social-material practices" (Barad 2007, p. 26). If we ignore the binaries because they are inseparable, entangled with each other through intra-action then ethics are part of that entanglement. Through agential realism, and the associated entangle-ment, cuts are made to produce phenomena, which are the mutual constituents of entangled agencies but these agencies are "only distinctive in a relational. . . sense" (Barad 2007, p. 33) and only distinctive through entanglement. Barad notes that because the possibilities for intra-acting "exist at every moment" "these changing possibilities entail an ethical obligation to intra-act responsibly in the world's becoming, to contest and rework what matters and what is excluded from matter-ing" (p. 178). As Barad notes in her interview with Rick Dolphijn and Iris van der Tuin:

> Ethics is about mattering, about taking account of the entangled materializations of which we are part, including new configurations, new subjectivities, new possibilities. Even the smallest cuts matter. Responsibility, then, is a matter of the ability to respond. Listening for the response of the other and an obligation to be responsive to the other, who is not entirely separate from what we call the self. This way of thinking ontology, epistemology, and ethics together makes for a world that is always already an ethical matter. ...ethics is not a concern we add to the questions of matter, but rather is the very nature of what it means to matter (Barad 2012, p. 69).

In other words, "cuts are agentially enacted, not by willful individuals but by the larger material arrangement of which "we" are a "part" (p. 178) and intra-actions and the associated cuts support the identification of together and apart, which is fluid. For Barad, ethics is about taking account of the entangled materialization acknowledging our responsibility for attending to the "tissue of ethicality that runs through the world" (p. 70).

10.4.1   Ethical Engagements

As an "accounting for how practices matter"Barad's diffractive methodology offers strategies for how we might consider ethical engagements. She acknowledges the material and semiotic contribution of Donna Haraway to the development of this methodology because Haraway proposed diffraction as an alternative to the thinking metaphor of reflection (Haraway 1992):

> Diffraction does not produce "the same" displaced, as reflection and refraction do. Diffraction is a mapping of interference, not of replication, reflection, or reproduction. A diffraction pattern does not map where differences appear, but rather maps where the effects of difference appear. Tropically, for the promises of monsters, the first invites the illusion of essential, fixed position, while the second trains us to a more subtle vision (Haraway 1992, p. 300).

For Haraway, the optical metaphor of reflection locked all participants into the idea that with reflection comes the expectation that when responding to an experience, the response will be a reflected image of the experience. However, diffraction allowed for "small but consequential differences" (p. 318) so such responses could show degrees of freedom in response not anticipated with reflection. Thus a variety of experiences would generate differences and for a researcher looking at a project through diffraction, there is an expectation of difference. Diffraction is comfortable with a heterogeneous history and there is no expectation that the phenomena, which we call learning, that emerges from participants' intra-actions with specific material aspects of the research project will be the same. Barad argues that knowing is a specific engagement with the world where a part of the world becomes differentially intelligible to other parts requiring a differential accounting to what matters and what is excluded from mattering. Haraway's diffraction metaphor disrupts the notion of representation and of reflecting back the represented reality. For Barad, diffraction is both phenomenon, a key discussion point in physics, and metaphor, for "describing a methodological approach" (Barad 2007, p. 71). For her, this methodology consists of reading insights by attending to, and being responsive to, the emergent details of "relations of difference" and their importance, "how they matter"(p. 71). It is in the differences that we detect learning. For us, this approach to research is very different from that which would be familiar to most researchers and teachers, even from a sociocultural stance, as they search for consistencies and coherences in data and, if differences are acknowledged, explore them as contradictions to the consistencies one might expect to observe within a culture.

Barad argues that diffraction is more profound than reflection because it is an "ethico-onto-epistemological matter" (Barad 2007, p. 381). By challenging the presumed and often accepted subject and object dichotomy, and other dichotomies, diffraction allows us to acknowledge that all humans intra-act with matter to relationally phenomenalize the world differently. For us, this means that in conducting research on teaching and learning. We have a responsibility to acknowledge and value the differential mattering of all participants, living and non-living. Baradian ethics is built upon the notion that one can affect others which includes non-human, and vice versa, culminating in the question "How to responsibly explore entanglements and the differences they make? (Barad 2007, p. 74). Barad's exploration of diffraction emphasizes how the substance and the diffraction grating impact the emergent phenomenon. We argue that Baradian-based ethics should:

1. Use a methodological approach that is performative (practices) and accepts that through intra-actions phenomena emerge rather than assuming preexisting separation of subject and object that requires representation to bring them into the same space.
2. Agential realism assumes that practices are associated with intra-acting that produces phenomena and uses an ontology that does not assume "words" and "things" and an epistemology that does not assume a correspondence between the two for truth.
3. Practices matter. Be respectful of entanglements of matter and humans recognizing that the specificity of entanglements is everything. We materialize the world differently through different practices.

Consequently, phenomena should be the focus of analysis rather than objects and subjects. Note that all these steps are performative, involving practices. For Barad it is through practice that a need for concepts and theory emerges. In the next section, we discuss the ethical implications for science education researchers when the nonhuman are considered, that is "who matters and what counts" (Taylor 2017, p. 5) combining a number of philosophical positions, including feminist post human or material feminism perspectives. Carol Taylor (2017) outlines how shifts from humanism to post humanism as a movement from a humanistic, anthropocentric focus on research to multi, inter, trans disciplinary theoretical approaches allows one to incorporate a plethora of theoretical frameworks, practices and methodologies. Concurrent with this explosion of theories and approaches is a shift in the subjectivities, relations, method/ology, and ethics. This posthuman world is populated by scholars across many disciplines including geography (Thrift 2008), anthropology (Malaforis and Renfrew 2010), cognitive science (Hutchins 2010) and education (Sørensen 2009) and they challenge us to explore ethics in this morethan-human world (Whatmore 2013) of material agency (Knappett and Malafouris 2008). Through her articulation of the ontology of agential realism, Barad brings together human and matter through intra-actions of "entangled agencies" through which "distinct agencies emerge" (Barad 2007, p. 33).

10.4.2 What Does Taking on a Baradian Perspective to Ethics 'Mean' for Science Education Research?

When science educators engage in research they are constantly making ethical choices and decisions. Heretofore though we argue that their decisions have not considered matter's agency, nor how humans (i.e. researchers students and participants) intra-act with materials. Consistently, science educators are making agential cuts in their research and teaching decisions, that ignore the material thus limiting the knowledge produced through research. To illustrate this point, we refer to Carol Taylor's study of how her ethical choice of engaging with Baradian theory where matter and meaning are not independent, but entangled and thus result in particular knowledge production. That knowledge production is situated and contextual (Haraway 1988). Specifically, she used Barad's concepts of entanglement, cut, phenomena and intra-action to examine the curriculum through diffraction of design, teaching and learning and knowledge production. In the following section we use Taylor's (2019) example of how she used Barad's agential realism to engage students' voice in framing, planning and crafting their curriculum, a critical examination of the teaching space and the knowledge produced to show how agential cuts generate differences, and those differences can matter. We use these three areas to illustrate what differences may emerge for science education research using an ethico-onto-epistemological framework.

10.4.3   The Ethics of Curriculum Choices

Taylor notes that in a Baradian context a curriculum 'comes into being' through the material-discursive engagements. The course Taylor discusses is arranged in modules and the first step is for the students to examine those modules and topics, and then discuss how they will organize the modules to develop the curriculum design. Through this practice, the students attain power and decide what matter matters in their coursework. This practice also disrupts the notion that learning has a progression. Could one teach science education in this open manner, providing students the opportunity to design the course? In a field where efforts in the past decade have focused on articulating and documenting the 'learning progression' of students' science knowledge. Are we ready? Will science educators consider that learning is not of a sequential nature but very context-bound? What are the ethical considerations to this approach?

What science would students choose to learn if we opened up this approach? Nearly a decade ago Gale Seiler and Allison Gonsalves (Seiler and Gonsalves 2010) reported on two urban high school science teachers who opened up the curriculum with the intent of engaging students in science. A commitment to Freirean principles underpinned the curricular reform and one goal was for the high school students to propose topics and ideas. Students expressed an interest in learning about the human body, however, the data reported show that most of the science topics were introduced into the discussion by the coteachers. The coteachers found sharing their power with students to establish the course curriculum to be challenging. After engaging students in conversations about what science would be learned and how it would be taught, the curriculum that was enacted reflected the teachers', not the students' choices. Students chose not to engage in the lessons and the 'curriculum' failed to achieve a critical goal of supporting students' science interests. In contrast, to students' reluctance to engage in teacher chosen curriculum, Seiler and Gonsalves

(2010) report the successful engagement of students in the dissection activities- a topic students identified that was of interest to them.

Students in Catherine Milne's environmental science course (Milne 2019a) explore how different intra-actions with different apparatus produce different phenomena. Informed by the work of Don Ihde (2015), students make various instruments and explore how each intra-action creates different forms of light phenomena. They note how moving from intra-action with a pinhole camera obscura, which they make, to glass prisms and then to spectroscopes establishes different forms of phenomena with respect to the nature of light and images. These intra-actions are also collective and relational and provide a context for other material discursive practices associated with observing, measuring, and asking questions. Milne challenges students to intra-act with glass prisms to form light spectra (See Fig. 10.1) and builds on that experience to explore the history of the discovery of infrared radiation by William Hershel (1802). Students then use Hershel's data, which he presented in tables, and construct ways of presenting the data not used by Hershel, such as graphs and diagrams. Since the course has a focus on environmental issues that includes global warming, the curriculum can use the experience with infrared radiation to explore energy transfer via convection, conduction and radiation and thermal equilibrium. Note the role of context, intra-action with matter and material-discursive practices and coherence in this segment of the curriculum.

Alexa, a student from Milne's course: *The instruments, thereby, certainly helped in observing the world differently. Before using these instruments, I had observed the spectrum, but had never analyzed the spectrum, considered how it forms, or contemplate[d] what would make it appear more visible.*

No one could accuse Alexa of being a materialist but her comment captures the intra-action between her and the various instruments she made as well as the way

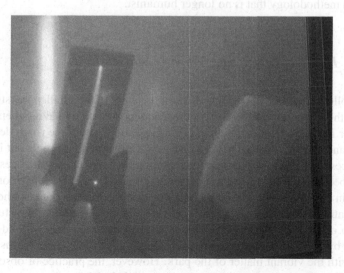

Fig. 10.1 Intra-action between student and prism with the goal of forming a light spectrum

the instrument intra-acted with the light. Note also that Milne made very definite choices in an attempt to develop a curriculum that is coherent where one experience builds on the next. So in this case, her approach is very different from that of Taylor. Also, the students in Milne's course were there, not because they loved science and wanted to major in a science with the goal of becoming a scientist. No, the courses fulfilled a required natural science prerequisite for their degrees whether their goals were to become an event planner, a music technologist, early childhood educator, or a food blogger.

Fundamental to re-thinking of ethics and science education research is the discussion about how our field would change if matter was taken seriously. For example, Elizabeth de Freitas and Anna Palmer (de Freitas and Palmer 2015) noted the significant impact on thinking about how people would learn science if one assumes "there is no ontological dualism between matter and meaning, or mind and body. Thus matter is conceptual, and concepts material. The implications for learning theory are tremendous, if complicated, as the focus turns to how children are entangled with concepts rather than merely engaged in (mis)recognizing them (p. 1203)." This poses a critical ethical question for science education, because IF science educators took matter seriously, then a large percentage of science education research needs to be re-thought, re-done, re-analyzed and possibly there would be new implications for the teaching and learning of science. As noted elsewhere, over 80% of science education research focuses on the conceptual learning of science (Chang et al. 2010) and there is no focus on how gender, ethnicity, sexuality and/or race impact that learning (Scantlebury and Martin 2010). This is a position argued also by Sørensen (2009) who notes that the biggest question any materialist approach to research in teaching and learning needs to address is "to account for how materials perform in school practices and what is performed" with such participation (p. 3). She notes also that from its position as a humanist endeavor education needs to develop a methodology that is no longer humanist.

10.4.4   Physical Space and Vibrant Materialities

As we follow Taylor's diffracting of the curriculum, she challenges students to attend to the "vibrant materialities' in the teaching context by having them identify the matter in the classroom and then posing questions about the room design, furnishings and the use of space at the university. One issue that stands out is the role of practices such as observing and questioning and how these practices are only possible because of the intra-action between the materials of the classroom and the students in Taylor's class. The same could be said for Milne's course where one of the first intra-actions for students is to go outside into the park where students are invited to observe the space, first with eyes open and then closed. Students were intrigued by the various phenomena that emerge from their various sensory intra-actions with the vibrant matter of the park. However, the practice of observation is only possible when matter and human are entangled and the phenomena produced is variable for all participants. In a previous study, Taylor (2016) noted how objects (a desk and chair), bodies (a male teacher) and space (classroom) became entangled and generated practices in the form of new questions about gendered pedagogical practices. The thing-power of clothing was also noted by Kathryn Scantlebury et al. (2018) and Kirsten Robbins (2016) when noting the changes in students' attitudes when donning lab coat or putting on the teacher's cardigan enacts changes in power and space.

What are the ethical questions science educators could raise by taking into consideration the physical spaces and the non-human material of when they are researching the teaching and learning science? What issues come to the fore when science is taught outside of a laboratory? For example, when teachers add things to their classes, such as posters, plants, and animals do they take into consideration the range of ethical concerns and choices this practice might generate?

In these examples, both Taylor and Milne identify moments in the curriculum that are diffracted through space and time, which Barad (2014) calls "spacetime-mattering" and these moments of intra-acting with spaces as "diffracted condensation" (p. 169), a "threading through" of entanglements, which for the students taught are never closed and never finished. Indeed, when reporting their learning in open class presentations many students in Milne's class identified their entanglement with the park and the emergent practice of observing as space-time co-ordinates that continued to haunt their ongoing entanglement with the course.

10.5   Material Impact of Tools and Instruments

If science educators: dismantle the dualisms between subject and object, ethics and knowledge; recognize that they are entangled in their research questions; AND in take into consideration how the material is impacting their work; we may collectively begin to identify differences that matter. Heretofore, most science education research has not taken into consideration the agentic nature of instruments and tools. For example, Milne (2019b, p. 13) argued that science educators often ignore the role of instruments in the teaching and learning of science. She posed the question, *"why is there so little attention given to understanding the role of instruments in the construction of knowledge?"* In exploring how humans/learners/students/teachers/researchers are entangled with instruments, science educators would acknowledge that we are 'being of the world', and as such constantly engaged in ethical decisions.

As Sørensen (2009) notes rarely in education research are different materials, used as interventions in a study, explored for how they change the forms of learning that takes place. This is especially the case in pre-test-post-test studies where the intervention is often seen as interchangeable. For example, consider a study exploring the learning of biology content comparing the impact of video versus lecture. Researchers assume that the learning of students assigned to one of those interventions can be compared using the same pre-test-post-test combination. Rarely do researchers explore how different materials may contribute to different forms of learning or problematize how different materials may change the types of interactions between teacher and student. However, shifting to using Barad's epist-ontological-ethical framework, challenges educators to ethical consideration of the influence of research tools, including pre-tests and post-tests, audio or video recorders, and other forms of technology such as smart tablets and phones or web based interfaces in educational contexts. For example, Nordstrom (2015) enacted a diffractive perspective to raise questions regarding the material-discursive practices of recording devices when conducting interview research. She noted how recording devices are used to make agential cuts, and raised questions about when and who decides to turn on a recording device. What material-discursive practices are enacted when this nonhuman entity becomes part of the research process? What differences come into being through the recording devices that researchers have ignored because they have not questioned the use or influence of the recording device in the research process (Nordstrom 2015). First, when used as a tool for conducting interviews, the recorder is a boundary making object, providing documentation of 'being there' in the research process and establishing the boundaries of an interview through when the recorder is turned on and off. Second, the ethical, political and personal considerations in using a recorder have rarely been problematized as its use is viewed as providing an unbiased documentation of the interview. Third, as yet many critical and post-modern theories have not problematized the recording device as objectivist-realist material-discursive practice. Specifically, Susan Nordstrom reported several ways in which recording devices disrupted her research such as (1) causing participants to feel ill at ease; (2) how a researcher's entanglement with the participant began before ethical steps such as obtaining permission to record were enacted; (3) and how extended interview/discussion maxed out the recorder's batteries limiting the documented data.

In another study, Cathrine Hasse (2019) discussed the influence of new technologies in Danish primary schools. Over several years (2011–2013) tablets were given to primary school children, the intent was to utilize technology to promote student learning in efficient ways. Concurrent with providing students with tablets, whiteboards replaced blackboards, and in some locales teachers, students and parents were confused as to what to do with the tablet. Moreover, Hasse (2019) notes that iPads were the tablet of choice because of Apple's large investments in Denmark. The introduction of iPads also had a noted unintended consequence, the iPads intra-acted with students' movement and exercise in ways that existed in direct opposition to the new Danish reforms aimed at getting children up and out exercising. Children become entangled with the iPads, which could be viewed as having what Bennett (2010) noted as 'thing-power' that is, is 'the curious ability of inanimate things to animate, to act, to produce effects dramatic and subtle' (Bennett 2010, p. 6).

Kathrin Otrel-Cass and Bronwen Cowie (Otrel-Cass and Cowie 2019) took a diffractive reading on their data from a project that focused on teacher-student interactions in science and technology classes and how materials and artifacts were incorporated into the teaching and assessment of student learning. Materials engaged students in exploring physical phenomena and when students' produced an artifact to provide evidence of their science learning, their ability to share their new knowledge improved. The teachers and researchers involved in this study, sought materials that the students connected with in ways that produced material-discursive practices that aligned with science knowledge.

Milne (2013) explored this issue with respect to the use of thermometers in the measuring of the boiling point of water and one of her regrets is how she did not problematize thermometers in her practice as a high school teacher of science when she had the chance to do so. We argue that appreciation of the agency of instruments could encourage all teachers to problematize instruments. So that, even with traditional confirmatory laboratory activities, teachers rather than directing students to explain why, when using a specific instrument, they did not get the expected results, which is typically how school science experiments are framed (see Millar 2004) being more diffractive and exploring further the phenomena produced as each student intra-acted with the laboratory activity.

10.6 Producing Knowledge

With respect to knowledge production, each iteration of a course is unique, materializing and stabilizing different forms of the course as entanglement produces anew. We share Taylor's argument that agential realism is generative for new modes of learning because it acknowledges the affective power of matter. In two courses, *Science in Our Lives: Environmental Science* and *Creativity Unbound,* Milne (2019a) challenges students to explore matter as evocative objects where the focus is on objects as companions in life experiences (Turkle 2007). Taylor also explores how agential realism collapses scale in experiences associated with learning, which we also find evocative. In Milne's course, zooming through different levels from the micro, "What role does a candle play in the production of carbon dioxide?", an exploration of how the chemical reaction of combustion contributes carbon dioxide gas to the local atmosphere, to the meso "What greenhouse gases do you contribute to the environment?", an exploration of how each student contributes to the carbon footprint based on the choices they make, to the macro "How do carbon dioxide emissions contribute to global warming?", which allows us to explore global systems and how human intra-action with industries contribute to the emission of greenhouse gases into the environment supports course participants to connect different levels of experience.

Taylor used autoethnography as a way for the students to document their engagement with matter, by noting their particular unique material-discursive practices and through those practices producing the knowledge they had learned during the course. This leads us into another ethical consideration – who decides which knowledge is of most worth? What areas of research have science educators ignored or chosen not to pursue?

Research and practice in science education have ignored post critical and post humanistic theories, remaining predominantly a conservative, heterosexual, white, masculine field (Lemke 2011). Science education researchers could use Barad and other post humanistic theories to examine indigenous science education. For example, Marc Higgins (2019) discusses the ethical considerations for science educators when deciding whether to include traditional ecological knowledge (TEK) and Indigenous ways-of-living-with-nature (IWLN) as well as Western modern science (WMS) in curriculum decisions. Using a Baradian framework, Higgins (2019) argues that there is not one ontology when science education seeks to 'trouble' the nature/culture binary and acknowledge that science knowledge becomes from *being in*, rather than from an outsider's perspective. He proposes that science education frames discussions on TEK and IWLN as an ontology which eliminates aligning these forms of science education with Cartesianism and provides the space for these knowledge forms to produce situated, relational knowledge.

Another example comes from Barad's statement (Barad 2007), "gender is a contested category whose intelligibility depends in part on the specifics of materializing structural relations… gender is constituted through class and community and other structural relations of power. Gender, class, and community are enfolded into, and produced through, one another (p. 243). In the 1980's, feminist science education researchers noted the similarity of practices by science teachers who had a 'track record' of encouraging girls in science (Kahle 1985). The teachers' classrooms were visually stimulating, they had posters that showed females and males engaged with science, often there were plants and animals in the room, providing a pleasant aesthetic. And in other ways, these teachers made their rooms 'like home'. Emphasizing the role of affect, for the girls, the science laboratory became a space where they could develop material-discursive practices that supported their science learning.

There are other results/outcomes from the science education studies that focused on gender that we can re-interpret by taking into consideration "specifics of materializing structural relations" and power dynamics. Fifield and Letts (2019) document the inhospitable climate within the science education research community for discussion and presentations of research using LGBT and queer theory as a framework. Their edited book with 17 chapters and two poems offers what Barad (2015) deemed new imaginings in queering science education knowledge across grade levels, formal and informal science contexts, pedagogical practices and curricular choices. Further research could engage with what Barad (2015) called for namely:

> What is needed is not a universalization of trans or queer experience stripped of all its specificities (as inflected through race, nationality, ethnicity, class, and other normalizing apparatuses of power), setting these terms up as concepts that float above the materiality of particular embodied experiences, but to make alliances with, to build on an already existing radical tradition (a genealogy going back at least to Marx) that troubles nature and its naturalness "all the way down." In doing so, it would be a mistake to neglect the spaces of political agency *within* science—its own deconstructive forces produce radical openings that may help us imagine not only new possibilities, new matter/realities, but also new understandings of the nature of change and its possibilities (Barad 2015, p. 413).

If science educators begin to consider a Baradian perspective on ethics, they could use a diffracted methodology to re-examine the field and look for the differences. In particular, we argue that a focus on the material-discursive practices that emerge when humans entangle with the non-human, the material, will produce different effects. Within agential realism, the ethical decisions made in conducting research are agential cuts. These cuts establish temporary boundaries around phenomena, but Barad challenges us to consider what differences are generated through these decisions and whether these differences matter. We argue that by ignoring the material, science education researchers have set boundaries and generated differences that matter both ethically and in the ways learners engage and entangle with the practice of science.

10.7   Finally...for Now

Research could be viewed as something one does, a place to collect information, or a practice involving intra-actions through which phenomena is emergent and in which we are changed in the process. Baradian theory involves understanding and using an epistemological-ontological-ethical framework that acknowledges the agency of matter, dissolves binaries because entities are entangled, and produces phenomena through agential cuts and context specific boundaries. This approach contrasts with modernity's ethical approaches of teleological, deontological and utilitarian ethics, which separate the human from matter, make humans the central and only agent and proceed to categorize humans into two groups, the researched and the researcher, that have differential access to power. In Baradian ethics, humans engage as entangled with a part in the world rather than existing at the top of a hierarchy of being. Barad and other feminist new materialism scholars challenge us in research to take into account matter and its impact within ethics. The use of Baradian-based ethics and a diffractive methodology that acknowledges the agency of matter would challenge science education researchers to produce new understandings-becoming (Barad 2014) with regards to curriculum, the agential cuts made when intra-acting to produce assignments, readings, course topics and assessments. Researchers are entangled in this work, and, as we have argued in this chapter, researchers are ethically bound to consider the consequences and inferences of this theory noting the differences produced and how those differences to the generation of knowledge in science education matter.

References

Barad, K. (2003). Posthumanist performativity: Toward an understanding of how matter comes to matter. *Signs: Journal of Women in Culture and Society, 28*(3), 801–831.

Barad, K. (2007). *Meeting the universe halfway: Quantum physics and the entanglement of matter and meaning.* London: Duke University Press.

Barad, K. (2012). Interview with Karen Barad. In I. Van der Tuin & R. Dolphijn (Eds.), *New materialism: Interviews & cartographies* (pp. 48–70). Ann Arbor: Open Humanities Press.

Barad, K. (2014). Diffracting diffraction: Cutting together-apart. *Parallax, 20*(3), 168–187. https://doi.org/10.1080/13534645.2014.927623.

Barad, K. (2015). TransMaterialities: Trans*/matter/realities and queer political imaginings. *GLQ: A Journal of Lesbian and Gay Studies, 21*(2–3), 387–422. https://doi.org/10.1215/10642684-2843239.

Bennett, J. (2010). *Vibrant matter: A political ecology of things*. Durham: Duke University Press.

Chang, D., Chang, D., & Tseng, K. (2010). Trends of science education research: An automatic content analysis. *Journal of Science Education and Technology, 19*, 315–331.

de Freitas, E., & Palmer, A. (2015). How scientific concepts come to matter in early childhood curriculum: Rethinking the concept of force. *Cultural Studies of Science Education, 10*, 1–22.

Fifield, S., & Letts, W. (2019). Queering science education without making too much sense. In W. Letts & S. Fifield (Eds.), *STEM of desire: Queer theories and science education*. Leiden: Brill|Sense Publishing.

Giddens, A. (1990). *The consequences of modernity*. Cambridge: Polity Press.

Haraway, D. (1988). Situated knowledges: The science question in feminism and the privilege of partial perspective. *Feminist Studies, 14*(3), 575–599.

Haraway, D. (1992). The promises of monsters: A regenerative politics for inappropriate/d others. In L. Grossberg, C. Nelson, & P. A. Treichler (Eds.), *Cultural studies* (pp. 295–337). New York: Routledge.

Hasse, C. (2019). Learning matter: The force of educational technologies in cultural ecologies. In C. Milne & K. Scantlebury (Eds.), *Material practice and materiality: Too long ignored in science education*. New York: Springer.

Higgins, M. (2019). Positing an(other) ontology: Towards different practices of ethical accountability within multicultural science education. In C. Milne & K. Scantlebury (Eds.), *Material practice and materiality: Too long ignored in science education*. New York: Springer.

Hutchins, E. (2010). Imagining the cognitive life of things. In L. Malafouris & C. Renfrew (Eds.), *The cognitive life of things: Recasting the boundaries of the mind* (pp. 91–101). Exeter: Short Run Press.

Ihde, D. (2015). Preface: Positiiong postphenomenology. In R. Rosenberger & P.-P. Verbeek (Eds.), *Postphenomenological investiogations: Eassays on human-technology relations* (pp. vii–xvi). London: Lexington Books.

Jackson, A. Y., & Mazzei, L. (2016). Thinking with a post human assemblage. In C. Taylor & C. Hughes (Eds.), *Posthuman research practices in education* (pp. 93–107). London: Palgrave Macmillan.

Kahle. J. B. (Ed.). (1985). *Women in science: A report from the field*. New York: Routledge Falmer.

Kant, I. (1785/1938). *The fundamental principles of the metaphysics of ethics*. New York: D. Appleton-Century.

Knappett, C., & Malafouris, L. (Eds.). (2008). *Material agency: Towards a non-anthropocentric approach*. New York: Springer.

Lemke, J. (2011). The secret identity of science education: Masculine and politically conservative? *Cultural Studies of Science Education, 6*, 287–292.

Lyotard, J.-F. (1984). *The postmodern condition*. Minneapolis: University of Minnesota Press.

Malaforis, L., & Renfrew, C. (2010). Introduction – The cognitive life of things: Archeology, material engagement and the extended mind. In L. Malafouris & C. Renfrew (Eds.), *The cognitive life of things: Recasting the boundaries of the mind* (pp. 1–12). Exeter: Short Run Press.

May, W. F. (1980). Doing ethics: The bearing of ethical theories on fieldwork. *Social Problems, 27*(3), 358–370.

Millar, R. (2004). *The role of practical work in the teaching and learning of science*. Paper prepared for the Committee on High School Science Laboratories: Role and vision. National Academy of Sciences, Washington DC.

Milne, C. (2011). Marie Curie and ethics in research. In M.-H. Chiu, P. Gilmer, & D. Treagust (Eds.), *Celebrating the 100th anniversary of Madam Maria Sklodowska Curie's Nobel prize in chemistry* (pp. 87–103). Dordrecht: Sense Brill Publishers.

Milne, C. (2013). *Creating stories from history of science to problematize scientific practice: A case study of boiling points, air pressure, and thermometers.* Conference paper for IHPST Meeting, Pittsburgh, June 19–22.

Milne, C. (2019a). Thinking about practices differently: Why materials matter. In C. Milne & K. Scantlebury (Eds.), *Material practice and materiality: Too long ignored in science education.* New York: Springer.

Milne, C. (2019b). The materiality of scientific instruments and why it might matter to science education. In C. Milne & K. Scantlebury (Eds.), *Material practice and materiality: Too long ignored in science education* (pp. 9–23). Cham, Switzerland: Springer Nature.

National Commission for the Protection of Human Subjects of Biomedical and Behavioral Research. (1978). *The Belmont report: Ethical principles and guidelines for the protection of human subjects of research* [DHEW Publication No. (OS) 78-0012]. Washington, DC: US Government Printing Service.

Nordstrom, S. N. (2015). Not so innocent anymore: Making recording devices matter in qualitative interviews. *Qualitative Inquiry, 11.* https://doi.org/10.1177/1077800414563804.

Otrel-Cass, K., & Cowie, B. (2019). The materiality of materials and artefacts used in science classrooms. In C. Milne & K. Scantlebury (Eds.), *Material practice and materiality: Too long ignored in science education.* New York: Springer.

Robbins, K. (2016). A matter of power. In N. Snaza, D. Sonu, S. Truman, & Z. Zaliwska (Eds.), *Pedagogical matters: New materialisms and curriculum studies* (pp. 153–163). New York: Peter Lang.

Rouse, J. (1991). Philosophy of science and the persistent narratives of modernity. *Studies in the History and Philosophy of Science, 22*(1), 141–162.

Scantlebury, K., & Martin, S. (2010). How does she know? Re-visioning conceptual change from feminist perspectives. In W. M. Roth (Ed.), *Re/structuring science education: Reuniting sociological and psychological perspectives* (pp. 173–186). New York: Springer. https://doi.org/10.1007/978-90-481-3996-5_12.

Scantlebury, K., Milne, C., & Hussenius, A. (2018). *Entangling matter and gender in the teaching and learning of chemistry part of symposium working across disciplines and differences for gender justice: Methodological, theoretical and practical challenges for feminist educators* paper presented at European education research conference. Italy: Bolzano.

Seiler, G., & Gonsalves, A. (2010). Student-powered science: Science education for and by African American students. *Equity & Excellence in Education, 43*(1), 88–104. https://doi.org/10.1080/10665680903489361.

Sørensen, E. (2009). *The materiality of learning: Technology and knowledge in educational practice.* New York: Cambridge University Press.

Taylor, C. (2016). Edu-crafting a cacophonous ecology: Posthumanism research practices for education. In C. Taylor & C. Hughes (Eds.), *Posthuman research practices in education* (pp. 5–24). London: Palgrave Macmillan.

Taylor, C. A. (2017). Rethinking the empirical in higher education: Post-qualitative inquiry as a less comfortable social science. *International Journal of Research & Method in Education, 40*(3), 311–324. https://doi.org/10.1080/1743727X.2016.1256984.

Taylor, C. A. (2019). Diffracting the curriculum: Putting 'new' material feminism to work to reconfigure knowledge-making practices in undergraduate higher education. In J. Huisman & M. Tight (Eds.), *Theory and method in higher education research* (Vol. 5). Bingley: Emerald Group Publishing Ltd.

Thrift, N. (2008). *Non-representational theory: Space| politics | affect.* London: Routledge.

Turkle, S. (2007). Introduction: The things that matter. In S. Turkle (Ed.), *Evocative objects: Things we think with* (pp. 3–10). Cambridge: MIT Press.

Whatmore, S. (2013). Political ecology in a more-than-human world: Rethinking 'natural' hazards. In K. Hastrup (Ed.), *Anthropology and nature* (pp. 79–95). London: Routledge.

Kathryn Scantlebury is a Professor in the Department of Chemistry and Biochemistry at the University of Delaware. Her research interests focus on feminist/gender issues in various aspects of science education, including post humanism, urban education, preservice teacher education, teachers' professional development, and academic career paths in academe. Scantlebury is a guest researcher at the Centre for Gender Research at Uppsala University, co-editor in Chief for Gender and Education, and co-editor of two book series with Brill Publishers.

Catherine Milne is a Professor in science education at New York University. Her research interests include urban science education, sociocultural elements of teaching and learning science, the role of the history of science in learning science, the development and use of multimedia for teaching and learning science, and models of teacher education. She is co-editor-in-chief for the journal, Cultural Studies of Science Education and co-editor of two book series for Springer Publishers and Brill Publishers.

Chapter 11
Students' Ethical Agency in Video Research

Jaakko Antero Hilppö and Reed Stevens

11.1 Introduction

The ability to record, replay and analyze human action has been at the forefront of many fundamental insights and discoveries within science education and the in the social sciences more broadly (e.g., Goldman 2007; Heath et al. 2010; Tiberghien and Sensevy 2012). Studies using video records as their primary data source have advanced our understanding, for example, of the different aspects of science education within classrooms, like teacher and student interaction or the nature of argumentation in science classroom (e.g., Osborne et al. 2004). Video records have also offered us insights into science learning in informal contexts, like science museums (Stevens and Hall 1997) or home (Hall and Schaverien 2001). Furthermore, video analysis has advanced our understanding of science instruction in different countries (Stigler et al. 2000) and how videos can be used as part of either pre-service or in-service science teacher education (Sund and Tillery 1969; Brophy 2003).

More recently, new ways of collecting and analyzing video data are becoming increasingly popular as participatory approaches spread within science educational research (i.e., Lundström 2013; Roberts 2011; Riecken et al. 2006; Rudman et al. 2017). While participatory methods have been part of the methodological tool kit of educational researchers for some time (Chambers 1994; Collier and Collier 1986), technological advances made in the last decade have significantly changed the

J. A. Hilppö (✉)
University of Helsinki, Helsinki, Finland
e-mail: jaakko.hilppo@helsinki.fi

R. Stevens
Northwestern University, Evanston, IL, USA

© Springer Nature Switzerland AG 2020
K. Otrel-Cass et al. (eds.), *Examining Ethics in Contemporary Science Education Research*, Cultural Studies of Science Education 20,
https://doi.org/10.1007/978-3-030-50921-7_11

extent of the researchers' work and the ways in which they can invite students and teachers to participate in research. Not only has the capacity and durability of different recorders increased exponentially, but at the same time their size, weight and price have significantly decreased as well. In effect, what previously would have been either highly implausible or even impossible to accomplish practically in terms of data collection, is now not only possible, but also available to a wide range of researchers and research groups. With standard, off-the-shelf consumer video equipment like action cameras (e.g., GoPro), recording students' and teachers' activities and reflections during an ongoing science project, how they engage with different forms of science in their everyday life or community action projects is now more feasible than ever before. In a similar fashion, the accessibility of different publicly available and shared video collections (like YouTube) has also created new opportunities for researchers to study phenomena like teaching and learning in new ways and from new data sets (Derry et al. 2010).

In addition to excitement, these new possibilities have also met with well-founded ethical reservations (e.g., Goldman 2007; Waller and Bitou 2011; Mok et al. 2015). The core argument of these warrants has been that while technological advances have made it possible to venture into unexplored sites, researchers should be reflexive about the possible ethical repercussions of these ventures and whether or not the potential gains of the studies outweigh their risks. In practical terms, although new video technology makes it possible for researchers to collect data on, for example, how children and youth engage with STEM topics and activities in the privacy of their own homes or rooms, providing them with cameras will also lead to breaching ethical boundaries. Possible scenarios include moments where small and silent recorders become invisible to participants who then do or say things on camera that they did not intend to share with the researchers or when a wearable camera records the participant's usernames and passwords when she or he works on a computer. In other words, more advanced and unnoticeable recorders might invade people's privacy and researchers should take this into account when designing their studies.

These cautions rightfully place the main burden of ethical deliberation on the researchers. Researchers' work often puts them in a privileged position. They can come to know things about the lives of their participants that are not common knowledge, and also potentially harmful if seen by others. In other words, taking part in research redefines the conventional boundary marking what people can know about each other in a way that accentuates the researchers' obligations of respectful and diligent treatment of this knowledge. Where this boundary goes is conventionally defined by the researcher who is more aware of the research process, its needs, possible outcomes and the impact that being involved might have on the participants' everyday lives.

At the same time, however, this emphasis draws attention away from the work that the participants themselves do to maintain and regulate this very same boundary. Although a research process officially starts after formal consent and assent have been given by the participants (i.e., children and their parents), whatever has been agreed on as being the scope of the study does not sustain itself automatically, but rather needs to be upheld along the way and renegotiated if breached. Furthermore, emphasizing the researcher's ethical agency also positions the participants and their competencies in a certain way, as in need of protection by others. More specifically in relation to video research, the cautions rest on a general assumption that during the research process the participants would not be aware of being recorded and, after becoming aware, could not act on this. In doing so, the cautions join and reinforce dominant narratives of participants', especially children's, vulnerability in research without empirically or conceptually exploring the credibility of their assumptions (e.g., Richards et al. 2015). If unchecked, the cautions could impede possible methodological advances in using video methods in research and unintentionally limit our understanding of teaching and learning in the STEM disciplines.

In this chapter, we engage with these assumptions in two ways. First, we draw on conceptualizations developed within the sociology of childhood, and especially in the literature concerning children's participation in research. In doing so, we outline how questions of children's competence come to define and negotiate the boundaries of the research process and how their own participation in research has been treated and conceptualized within this literature. Second, we share demonstrative vignettes from our own ethnographic work that show how children indicate their awareness of the audience of nearby recorders and, through such actions, also create spaces for private, out-of-view interaction they do not wish to be recorded. Through our vignettes, we broaden the scope of Christensen's and Prout's (Christensen and Prout 2002) notion of ethical symmetry.

11.2 Working Towards a Fuller Ethical Symmetry

Children's participation in research has been a central topic within the sociology of childhood literature (Christensen 2004; Gallacher and Gallagher 2008). For this broader body of work, the way and extent to which children are asked and allowed to take part in research activities is an important site where different societal perspectives on childhood come to fore, a core interest for childhood sociology. Within this literature, there have been many attempts to come to terms with how the "messiness" of children's participation in research should be treated in ethical ways (e.g., Punch 2002; Komulainen 2007). In other words, researchers have tried to both conceptually and practically deal with situations in which children, for example, express their willingness to be part of the research in multiple different ways, change their mind throughout the research process or change what they want to share with the researchers. The challenge has been to find means that allow researchers to continue working with children in ways that ethically accommodate and engage with this heterogeneity of participation.

The notion of *ethical symmetry*, introduced by Christensen and Prout (2002), was one of the first steps in this process. In their article, after first discussing how the then new perspective of "children as social actors" was impacting the field of social sciences, Christensen and Prout suggested that the best way for researchers to accommodate children's different competencies to be part of the research process, without curtailing their abilities to do so, was to start with the assumption that they were as capable as adults. In other words, by ethical symmetry they mean *"that the researcher takes as his or her starting point the view that the ethical relationship between researcher and informant is the same whether he or she conducts research with adults or with children."* (Christensen and Prout 2002, p. 48). For Christensen and Prout, this symmetry is the starting point for a dialogue between researchers and the participating children about their ways of being part of the research. That is, the researcher and her or his practices should evolve during the research process to achieve a moment-to-moment goodness-of-fit ethically. In addition to this practical orientation, Christensen and Prout also argue that the principle of ethical symmetry should also be seen as a value-based choice that serves to develop and guide shared ethical grounds and guidelines among researchers, within their specific field as well as the scientific community overall.

After Christensen's and Prout's work, other similar relational and process-oriented notions of research ethics with children have been introduced. One such notion is *process assent*. This has been used to highlight how during the research children's wills and wants regarding how, what and when they wish to share their lives with researchers change and the fact that the researcher needs to be reflexive about this before and during the research process (e.g., Flewitt 2005; Alderson 2005). In a recent article on children's assent, Dockett et al. (2013) explain that:

> As with consent, providing assent can be an ongoing process, with the decision to partici-
> pate, or not, renegotiated or revoked at any time (Cocks 2007). This approach is referred to
> as process assent (Alderson 2005; Cutliffe amp; Ramcharan 2002; Flewitt 2005) as it
> involves the renegotiation of assent over the life of the research, as new information is pro-
> vided or new data are generated. (p. 3)

What the notion of process assent helps highlight and conceptualize is the process of assent beyond the start of the research. As such, it provides for a way to orient to and describe the ways in which children might oscillate between wanting, or not, to be part of the research and negotiating the boundaries of this participation with the researchers.

While both ethical symmetry and process assent have been important contributions to discussions about research ethics with children, their use has been largely limited to situations where the boundaries of the research are negotiated between children and adults (although cf., Dockett et al. 2013; Christofides et al. 2016). That is, the way in which possible issues of assent, privacy and the boundaries of the research work are handled by children themselves within their peer interactions have not been in focus of previous work. Within studies that employ participatory methods with children, the possibility and importance of such situations are often acknowledged and researchers like Mary Kellett (2005) offer guidance on how to teach children to do ethical research. Occasionally researchers also share narrative examples from their field work on how the participating children have handled such issues or situations. For example, Hilppö (2016), when discussing his own co-participatory studies in which children were asked to document their everyday life with cameras, noted that:

> Although the need to be respectful of others was emphasized and discussed with the pupils, as was making an effort to frame the photographs so that only consenting persons were shown, the mere fact of pupils taking photographs and looking at them in joint school spaces – something which is not a common practice – created disruptions. On occasion, the participating pupils' playful orientation to the documentation and taking of mock-up shots made some teachers as well as other pupils wary of the research and dubious of whether sufficient attention had been given to the issue of privacy. (p. 32–33)

While such narrative examples are illuminating in themselves, the way in which defining, regulating and maintaining the boundaries of the research efforts are done by the children themselves within their peer interactions have not been in focus of previous work, conceptually or empirically. In the remainder of this chapter, we present vignettes from our own ethnographic work that focus precisely on this issue. In these vignettes, students indicate their awareness of the audience of nearby recorders to each other and how, through such actions, they also create spaces for private, out-of-view interactions they do not wish to be recorded. As such, the vignettes engage with Christensen's and Prout's (2002) notion of ethical symmetry by showing how its scope can be extended to include children's peer negotiations about the boundaries of the research activities.

11.3   Student's Ethical Agency with FUSE

The demonstrative vignettes we share and analyze below come from a yearlong ethnographic investigation into student learning and student experiences of an alternative learning infrastructure called the FUSE Studio (Stevens et al. 2016; Ramey and Stevens 2018). The FUSE Studio is designed to act as an on-ramp for students' interest, discovery and development in science, technology, engineering, arts and mathematics, or STEAM. In addition, the FUSE Studio model also aims at developing students' collaboration skills, creativity, critical thinking and other connected competencies often associated with the broad notion of "twenty-first century skills". The core activities of FUSE revolve around a suite of 25 different STEAM challenges that students are freely allowed to choose from and complete at their own pace. The challenges range from building solar cars, laser mazes and roller coasters to 3D printing jewelry, writing code for video games and designing houses with 3D modelling software.

During the academic year 2015–2016 we collected data in seven different FUSE Studio implementations in three different schools located in a large midwestern school district in the United States. Each Studio did FUSE for 90 min a week for the whole year as part of their fifth and sixth grade science curriculum. In the beginning of each session, we asked seven students if they would wear a visor camera, an action camera attached to a visor cap (see Image 11.1), while they worked. Wearing a visor camera each time depended on the students' willingness to do so. That is, in

Image 11.1 Student wearing a visor camera

addition to formally assenting to being part of the research process in the beginning of the school year, wearing the visor camera was not mandated by the research design, but rather was an opt-in feature of the design. In the beginning of the study, we informed the students that only we, the researchers, would see what was captured on the visor cameras and that our intention was to understand what they did and learned while being in the studio. We also explained that the camera would allow us to follow their learning from their perspective much closer than a long shot camera in the back of the classroom.

During the study, when the students wore the visor camera, we did not regulate or try to control what they did with the cameras, apart from occasionally reminding them to keep the camera with them (if they had taken it off) and making sure that they were recording. Overall, the students' orientation to the cameras changed during the year, shifting gradually from an enthusiastic uptake of the cameras being a standard part of the studio materials with some students to lack of enthusiasm and refusals to wear the camera with others. Some students also gradually moved form a more reserved orientation to the cameras to wanting to wear them toward the end of the year. From time to time the students would play with the cameras by making faces at them and talking directly to them, but mostly, they did not pay special attention to the cameras.

During the data collection phase, however, we noticed moments between the students where they turned their visor cameras away from some part of their interactions for a short moment but, importantly, continued to use the visors to record their interactions afterwards. In other words, the students seemed to display an intention to limit what they wanted to share with us researchers, but also a commitment to the research process in general. After the school year had ended, to investigate these interactions further, we searched through the content logs (Jordan and Henderson 1995) we had produced during the year for instances where the visor camera was explicitly mentioned as being the focus of the participants. Of the 360 identified episodes most dealt with instances where the camera was handed by the researcher to the participating student or from the student back to the researchers. Because these situations flee outside the scope of our interest, we further selected only situations where the students were interacting together without adult presence and ended up with 45 episodes altogether. In addition to situations where the students played around with the visor camera, our interactional analysis revealed situations where students indicated to each other their awareness of the audience of nearby recorders and how, through such actions, they also created spaces for private, out-of-view interactions they wished not to be recorded. Below, we share two illuminating vignettes of such interactions.

11.3.1 *"I Can't. I Have Visor Camera"*

In the first vignette, two sixth-grade students, Tamaz and Nuri, were working side-by-side on different challenges on adjacent computer stations. Tamaz was wearing a visor camera. He was designing his dream home with an AutoCAD software called SketchUp, when Nuri turned to him and asked:

Turn	Student	Verbal action
1	Nuri	Tamaz?
2	Tamaz:	Yeah?
3	Nuri:	We are not doing this?
4	Tamaz:	I can't. I have visor camera (whispered).
5	Nuri:	You had it last time didn't you?
6	Tamaz:	No
7	Nuri:	Well it's ok.
8	Tamaz:	We'll do it after. Well, I still need to add the arcade machine into my house.

What took place in this brief episode was a discussion between Tamaz and Nuri about doing something they called "this". While what in specific "this" referred to did not get revealed during the interaction, the way in which Tamaz and Nuri treat the suggestion reveals that doing "this" was problematic because of the visor camera. After a brief discussion about the matter Tamaz and Nuri decide to continue with what they have been doing and postpone Nuri's suggestion to a later time.

In this vignette, students' ethical agency was present in two ways. First, the way in which Tamaz drew Nuri's attention to his visor camera (Turn 4) indicates that Tamaz treated Nuri's suggestion as problematic and in need of being handled somehow. By topicalizing the visor camera and lowering his voice, Tamaz positioned doing "this" as something that should not be captured on the camera. In effect, Tamaz's suggestion opened up a negotiation about the boundary between what should and should not be shared with the researchers and to which side doing "this" belongs to. Tamaz's suggestion also positions doing "this" as being different in this sense from what the students were doing. Second, after Nuri did not initially agree with his positioning of the visor camera as problematic, Tamaz suggested a new way to handle the issue (Turn 8), doing "this" later, probably when Tamaz is not wearing the visor camera. Although Nuri did not explicitly respond to Tamaz, the fact that both of them continue with what they were doing before tells us that Nuri agreed with Tamaz. What, in other words, Tamaz and Nuri accomplish with their interaction is to negotiate the boundary between their personal lives and what they want to share with the researchers.

11.3.2 Emil Re-positions His Visor

In the second vignette, Emil, a sixth-grade boy wearing the camera, was doing a challenge by himself. Jaden, Amali and Dereck were working next to Emil on their own challenges. While Emil often agreed to wear the visor camera, the three other students had occasionally expressed not wanting to be recorded directly, although they had assented to being part of the research. In the vignette Emil, who frequently collaborated with the three students, acted according to their wishes by turning his visor away from them after realizing its direction. In contrast to the previous episode, in this situation the maintenance of the boundaries of the research is done by physical action alone. Unlike, with Tamaz and Nuri, Emil re-positions his visor camera without negotiating about it explicitly with the other students.

During the studio session Emil had laid the visor camera down on the table, something that the students did from time to time. He had positioned the camera so that it recorded him working on the computer, but the camera was also directed towards Jaden, Amali and Derek. While Emil was working on his challenge, the following interaction took place (Image 11.2).

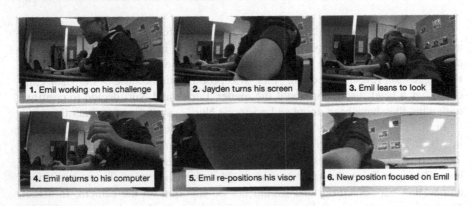

Image 11.2 Emil re-positioning his visor camera

Turn	Student	Verbal action	Nonverbal action	
1	Emil:		Beatboxes while working on the challenge and laughs to himself	Image 1 in Image 11.2
2	Jaden:	(unclear)	Turns his screen toward Amali and Derek	Image 2 in Image 11.2
3		(Amali and Derek laugh)		
4	Emil:		Turns his head toward Jaden and leans over	Image 3 in Image 11.2
5	Emil:	What does that Kong do?		
6		Emil, Amali and Jaden laugh		
7	Emil:		Returns to his screen and turns the camera more toward himself	Image 4 and 5 in Image 11.2
8	Emil:		Continues working on his challenge	Image 6 in Image 11.2

The episode begun when Emil was working on his challenge, beatboxing and laughing at something on his screen (turn 1). Next, Jaden turned his screen towards the two other boys, Amali and Derek, who laughed at what was on Jaden's screen (turns 2 and 3). Emil then turned to look and leaned in to see what was on Jayden's computer screen (turn 4). After asking about what he sees on the screen and laughing with Amali and Jaden (turns 5 and 6), Emil returns back to face his own computer screen. Importantly, when returning Emil looks at the camera and turns it away from Jaden, Amali and Derek, but so that it still captures what he is doing (turn 7). After this, Emil continues working on the challenge (turn 8).

Students' ethical agency is present in this second vignette in much the same way as in the first one. Like Tamaz and Nuri, Emil draws a boundary between what the researchers can and cannot see, in this case by re-positioning the camera. However, on this occasion the boundary is redrawn after Emil has realized that his visor camera was directed at the boys. Importantly, Emil's decision to re-position the camera presents a moment in which Emil takes up the responsibility of safeguarding the other students from the view of the camera and hence maintaining the boundary between what is and is not shared with the researchers.

11.4 Discussion

Video records of human action have been a valuable resource for producing many of the fundamental insights and discoveries not just in education and specifically in STEM education but also more broadly in the social sciences (e.g., Heath et al. 2010). Through many technological advances, our abilities to record and analyze human interaction have dramatically increased over the recent decades (Downing and Tenney 2008) and these developments are redrawing the ethical boundaries of research work (e.g., Mok et al. 2015). In this chapter we have shown, drawing on our own ethnographic video research in an alternative learning infrastructure, the FUSE Studio, how students manage the boundary between what the researchers are and are not allowed to know about their own lives and peer interactions. By highlighting moments of students' ethical agency, we have shown that students not only are aware of the presence of the video recorders and what they are recording, but also balance their commitment to the data collection and their own and other students' personal relation to the cameras. Through this work, we have shown how the notion of *ethical symmetry* (Christensen and Prout 2002) can be extended to cover the work that children do among themselves to manage the boundaries of the research. More specifically, our work shows that there is symmetry between the researcher's and the participants' positions in this regard.

Discussions around the ethics of children's participation in research often oscillate between positions that argue for children's vulnerability (and subsequent need of protection) or for their capability and autonomy to participate. Within these discussions, it is often acknowledged that while institutional safeguards, like review boards, mandatory ethics courses and guidelines are crucial in protecting all parties to the research process and making commitments transparent, at the same time they do not accurately represent the ethical deliberation process on the ground (e.g., Sleeboom-Faulkner et al. 2017). For example, the ways in which access to the research site is negotiated or how consent and assent are often acquired from the participants speak for a relational ethical position that allows to conceptualize the research work as a living–not static–process (Dockett et al. 2013; Christofides et al. 2016; Hilppö et al. 2019). If interpreted too strictly, these safeguards can also overreach their protective agenda and impose limitations on the research that hamper the advancement of the field, especially when technological advancements, like with video technology, offer new avenues for the research to explore. Importantly, such overreach also runs the ethical risk of misrepresenting and treating the participants of the research as incapable of weighing the risks of participation and regulating what they share with the researchers themselves. In relation to these arguments, the symmetry we have argued for in this chapter aligns with recent calls that question such assumptions (e.g., Richards et al. 2015) and reiterates the need to conceptualize and present the relationship between the researchers and the participants of the research, also as complex processes.

One possible way to open up and present this complexity could be sharing case narratives of how the boundaries of the research process have been negotiated and managed throughout the life time of a project by the researchers and, for example, between them and a review board. Being transparent about these negotiations would importantly bring to fore the division of labor between the parties in practice and their contribution to securing the ethicality of the research. As such, documenting these processes and sharing them in narrative form would be one way conducting and demonstrating the aforementioned relational ethics in practice. As a practice, such documentation would also easily align with notions like process assent presented earlier in this chapter. These narratives, and the transparency they offer, would also be an important resource for teaching and learning about research ethics and crucially how the ethicality of the research practice is secured when the research project is reaching out to new avenues of research with new methods and technology (cf., Pyyry 2012; Allen and Israel 2018).

Our contribution highlights that an important ingredient of such narratives could also be the ethical agency of the participants, and especially how the participants themselves regulate where the boundaries of the research are. While different ethical safeguards are needed, the boundaries they establish do not maintain themselves and part of that upkeep is done by participants. Highlighting these moments in themselves, as we have done here, or within the overall narrative of the research process could possibly demonstrate how the negotiation of the ethics of the research has not been solely the domain of the researchers. Importantly, such moments represent significant opportunities for us researchers to be more reflexive, analytical and transparent about the ethics of our work. In this vein, our contribution encourages both researchers and gatekeepers, like review boards, to be analytical, not conjectural, when weighing the risks and potential impact of new research technologies, like new forms of video search. That is, identifying, analyzing and reporting moments where the boundaries of the research are explicitly negotiated offer significant avenues for us researchers to be transparent about the ethics of our work and also of our participants understandings of them. By this, our contribution highlights that analyzing such interactions creates opportunities for being reflexive not only about the validity of our methodological choices (Speer and Hutchby 2003; Hilppö et al. 2017), but also about their axiology in action.

References

Alderson, P. (2005). Ethics. In A. Farrell (Ed.), *Ethical research with children* (pp. 27–36). New York: Open University Press.

Allen, G., & Israel, M. (2018). Moving beyond regulatory compliance: Building institutional support for ethical reflection in research. In R. Iphofen & M. Tolich (Eds.), *The SAGE handbook of qualitative research ethics* (pp. 453–462). London: Sage.

Brophy, J. (Ed.). (2003). *Using video in teacher education*. Bingley: Emerald Group Publishing Limited.

Chambers, R. (1994). The origins and practice of participatory rural appraisal. *World Development, 22*(7), 953–969.

Christensen, P. H. (2004). Children's participation in ethnographic research: Issues of power and representation. *Children & Society, 18*(2), 165–176.

Christensen, P., & Prout, A. (2002). Working with ethical symmetry in social research with children. *Childhood, 9*(4), 477–497.

Christofides, E., Dobson, J. A., Solomon, M., Waters, V., & O'Doherty, K. C. (2016). Heuristic decision-making about research participation in children with cystic fibrosis. *Social Science & Medicine, 162*, 32–40.

Cocks, A. (2007). The ethical maze: Finding an inclusive path towards gaining children's agreement to participation. *Childhood, 13*(2) 247–66.

Collier, J., & Collier, M. (1986). *Visual anthropology: Photography as a research method*. Albuquerque: UNM Press.

Cutliffe, J.R., & Ramcharan, P. (2002). Levelling the playing field? Exploring the merits of the ethics-as-process approach for judging qualitative research proposals. *Qualitative Health Research, 12*(7): 1000–10.

Derry, S. J., Pea, R. D., Barron, B., Engle, R. A., Erickson, F., Goldman, R., et al. (2010). Conducting video research in the learning sciences: Guidance on selection, analysis, technology, and ethics. *The Journal of the Learning Sciences, 19*(1), 3–53.

Dockett, S., Perry, B., & Kearney, E. (2013). Promoting children's informed assent in research participation. *International Journal of Qualitative Studies in Education, 26*(7), 802–828.

Downing, M. J., & Tenney, L. J. (Eds.). (2008). *Video vision: Changing the culture of social science research*. Newcastle: Cambridge Scholars Publishing.

Flewitt, R. (2005). Conducting research with young children: Some ethical considerations. *Early Child Development and Care, 175*(6), 553–565.

Gallacher, L. A., & Gallagher, M. (2008). Methodological immaturity in childhood research? Thinking through participatory methods. *Childhood, 15*(4), 499–516.

Goldman, R. (2007). Video representations and the perspectivity framework: Epistemology, ethnography, evaluation, and ethics. In R. Goldman, R. Pea, B. Barron, & S. J. Derry (Eds.), *Video research in the learning sciences* (pp. 3–37). New York: Routledge.

Hall, R. L., & Schaverien, L. (2001). Families' engagement with young children's science and technology learning at home. *Science Education, 85*(4), 454–481.

Heath, C., Hindmarsh, J., & Luff, P. (2010). *Video in qualitative research. Analysing social interaction in everyday life*. London: SAGE Publications Ltd.

Hilppö, J. (2016). *Children's sense of agency – A co-participatory investigation*. Doctoral thesis. University of Helsinki.

Hilppö, J., Lipponen, L., Kumpulainen, K., & Rajala, A. (2017). Visual tools as mediational means: A methodological investigation. *Journal of Early Childhood Research, 15*(4), 359–373.

Hilppö, J., Chimirri, N. A., & Rajala, A. (2019). Theorizing research ethics for the study of psychological phenomena from within relational everyday life. *Human Arenas, 2*(4), 405–415.

Jordan, B., & Henderson, A. (1995). Interaction analysis: Foundations and practice. *The Journal of the Learning Sciences, 4*(1), 39–103.

Kellett, M. (2005). *How to develop children as researchers: A step by step guide to teaching the research process*. London: Sage.

Komulainen, S. (2007). The ambiguity of the child's 'voice' in social research. *Childhood, 14*(1), 11–28.

Lundström, M. (2013). Using video diaries in studies concerning scientific literacy. *Electronic Journal of Science Education, 17*(3).

Mok, T. M., Cornish, F., & Tarr, J. (2015). Too much information: Visual research ethics in the age of wearable cameras. *Integrative Psychological and Behavioral Science, 49*(2), 309–322.

Osborne, J., Erduran, S., & Simon, S. (2004). Enhancing the quality of argumentation in school science. *Journal of Research in Science Teaching, 41*(10), 994–1020.

Punch, S. (2002). Research with children: The same or different from research with adults? *Childhood, 9*(3), 321–341.

Pyyry, N. (2012). Nuorten osallisuus tutkimuksessa. *Menetelmällisiä kysymyksiä ja vastausyrityksiä. Nuorisotutkimus, 1*(2012), 35–53. [Title in english: Youth participation in research. Methodological questions and possible answers.].

Ramey, K. E., & Stevens, R. (2018). Interest development and learning in choice-based, in-school, making activities: The case of a 3D printer. *Learning, Culture and Social Interaction*.

Richards, S., Clark, J., & Boggis, A. (2015). *Ethical research with children: Untold narratives and taboos*. Basingstoke: Springer.

Riecken, T., Conibear, F., Michel, C., Lyall, J., Scott, T., Tanaka, M., et al. (2006). Resistance through re-presenting culture: Aboriginal student filmmakers and a participatory action research project on health and wellness. *Canadian Journal of Education/Revue canadienne de l'éducation, 29*, 265–286.

Roberts, J. (2011). Video diaries: A tool to investigate sustainability-related learning in threshold spaces. *Environmental Education Research, 17*(5), 675–688.

Rudman, H., Bailey-Ross, C., Kendal, J., Mursic, Z., Lloyd, A., Ross, B., & Kendal, R. L. (2017). Multidisciplinary exhibit design in a Science Centre: A participatory action research approach. *Educational Action Research, 26*, 1–22.

Sleeboom-Faulkner, M., Simpson, B., Burgos-Martinez, E., & McMurray, J. (2017). The formalization of social-science research ethics: How did we get there? *HAU: Journal of Ethnographic Theory, 7*(1), 71–79.

Speer, S., & Hutchby, I. (2003). From ethics to analytics: Aspects of participants' orientations to the presence and relevance of recording devices. *Sociology, 37*(2), 315–337.

Stevens, R., & Hall, R. (1997). Seeing Tornado: How VideoTraces mediate visitor understandings of (natural?) spectacles in a science museum. *Science Education, 18*(6), 735–748.

Stevens, R., Jona, K., Penney, L., Champion, D., Ramey, K. E., Hilppö, J., Echevarria, R., & Penuel, W. (2016). FUSE: An alternative infrastructure for empowering learners in schools. In C. K. Looi, J. L. Polman, U. Cress, & P. Reimann (Eds.), *Transforming learning, empowering learners: The International Conference of the Learning Sciences (ICLS) 2016, Volume 2* (pp. 1025–1032). Singapore: International Conference of the Learning Sciences.

Stigler, J. W., Gallimore, R., & Hiebert, J. (2000). Using video surveys to compare classrooms and teaching across cultures: Examples and lessons from the TIMSS video studies. *Educational Psychologist, 35*(2), 87–100.

Sund, R. B., & Tillery, B. W. (1969). The use of the protable television tape recorder in science education. *Science Education, 53*(5), 417–420.

Tiberghien, A., & Sensevy, G. (2012). The nature of video studies in science education: Analysis of teaching & learning processes. In D. Jorde & J. Dillon (Eds.), *Science education research and practice in Europe: Retrospective and prospective* (pp. 140–179). Rotterdam: SensePublishers.

Waller, T., & Bitou, A. (2011). Research with children: Three challenges for participatory research in early childhood. *European Early Childhood Education Research Journal, 19*(1), 5–20.

Jaakko Antero Hilppö is a postdoctoral researcher at the Faculty of Educational Sciences, University of Helsinki. His research has focused on children's sense of agency in educational institutions and everyday life and co-participatory research methods with children. He has also studied compassion in children's peer interactions and cultures of compassion in early childhood and care contexts. More recently, Hilppö's attention has focused on children's projects as manifestations of their agency in and across different settings and especially on compassionate projects and the learning taking place within them. Hilppö is also a lazy high baritone and has watched C-beams glitter in the dark near the Tannhäuser Gate.

Reed Stevens is a Professor of Learning Sciences in the School of Education and Social Policy at Northwestern University and the founder of FUSE Studios, an educational experience for both school and out of school environments (https://www.fusestudio.net/). His research focuses on learning in and across informal and formal settings, particularly those that are connected to STEAM ideas and disciplines. Since the early 1990s, he has conducted ethnographic studies of cognition, creativity, and learning in settings that include homes, science museums, K-12 STEM classrooms, early childhood learning centers, undergraduate engineering education, and a range of professional STEAM workplaces. His research program has the broad goal of building an understanding of learning across the lifespan in everyday life and designing new kinds of learning experiences from the insights of these ethnographic studies. He has expertise in a broad range of ethnographic field methods for studying cognition and learning, with a specialization in techniques for analyzing moment-to-moment interaction between people and with objects and technologies. He co-led two NSF Centers (Learning in Informal and Formal Environment [LIFE] and the Center for Advancement of Engineering Education [CAEE]) while a professor at the University of Washington.

Chapter 12
The Performativity of Ethics in Visual Science Education Research: Using a Material Ethics Approach

Kathrin Otrel-Cass

12.1 Better Data for Science Education Research

Education research is the scientific field of study that examines education and learning processes and the human attributes, interactions, organizations, and institutions that shape educational outcomes. Scholarship in the field seeks to describe, understand, and explain how learning takes place throughout a person's life and how formal and informal contexts of education affect all forms of learning. Education research embraces the full spectrum of rigorous methods appropriate to the questions being asked and also drives the development of new tools and methods. (American Education Research Association, n.d.)

The American Education Research Association provides fairly broad descriptions on the aims and goals of education research. To further refine this in the area of science education research it is easiest to just 'add science'. While science education is just as important as any other learning area, it is a subject that is highly political and attributed with great economic consequences and attracts much attention amongst those concerned with societal economic prosperity. This means that science education research that support or perhaps demand changes in science teaching or learning is of significant consequence. Osborne and Dillon (2008) remind their readers about the need for more critical approaches in science education research, for instance to appropriately address what kind of science competences are in fact needed in the future. They write that "… better data is needed before making major policy decisions on science education" since "persuading young people to pursue careers in science without the evidence of demand would be morally questionable" (2008, p.7). The question is, however, what could possibly constitute better data? How do we perceive 'better data' to include so it can deal with addressing popular science education research questions and be ethically justifiable?

K. Otrel-Cass (✉)
University of Graz, Graz, Austria
e-mail: kathrin.otrel-cass@uni-graz.at

© Springer Nature Switzerland AG 2020
K. Otrel-Cass et al. (eds.), *Examining Ethics in Contemporary Science Education Research*, Cultural Studies of Science Education 20,
https://doi.org/10.1007/978-3-030-50921-7_12

When researchers present and discuss their research ideas and findings it also shines light on the ethical decisions that were made as part of conducting research projects. The decisions on how data has been collected and analysed can touch upon interdisciplinary fields, including sociology, philosophy, and psychology, each with their nested ontologies and epistemologies. However, this ethical perspective is rarely elaborated on when research is presented (Burgess 2005). The traditional modus of research production (conference presentations, articles etc.) includes a degree of repetitiveness (including that of methods or research designs). It can be said that research is therefore dealing with aspects of performativity or "ritualized production" (Butler 2011, p.60). This ritualisation carries the problem that researchers may become blind to ethical decision making and the impact of their research on people.

In the case of classroom-based science education research, it is quite common that researchers are focused at the micro level of research, that could include interactions that unfold between students and teachers and the concrete challenge can be that if not careful, a researcher can lose his or her appreciation on how micro level aspects of their analysis might cause harm to their participants. While ethics guidelines ought to steer researchers against this kind of blindness of impact (see also BERA 2018) it is not always achieved.

I want to focus in particular on visual and video-based research at the micro level (with focus on individual/s experiences, classroom activities or specific teaching practices) since these micro details shed light back on the meso and macro level of science education. For instance, implementations coming from policy makers (macro level) can be witnessed and traced to see how they are implemented back in the classroom environment (meso level) and enacted in interactional situations (micro level) (Roth et al. 2008). Working at the micro level is often about the analysis of what is visible during observations of science education practices and what is materialised in notes, imagery, and/or videography. The specific point here is that each data set documents the negotiations of what has been agreed on to be captured between researchers and participants. The complexity that visual data presents means also that they are always in some way joint products (i.e. students and teachers perform in front of a researcher's camera, photos of people or materials capture specific acts).

The ethics of conducting science education research with human participants presupposes the existence of certain sets of values (for both the researcher and participants) and those values are not necessarily shared. Kelly (2011, p.4) writes in a review about the philosopher Max Scheler's ideas on ethics that "values become functional as they guide action" and that "values exist only when they function in the 'performance' of the living spirit in human thought and action." The notion of performativity in ethics implies that ethical practices are enacted and subject to change. This stance suggests the need for responsiveness and reactiveness to the different settings and situations when "awareness of meanings and values becomes active in the thought and the behaviour of individuals and groups" (Kelly 2011, p.4). Kelly notes also that "...Scheler believed that we need sociology of knowledge adjunctive to ethics to study how values function in the moral consciousness of cultures and persons, and how these evolve" (2011, p.5).

This chapter focuses particularly on science education research dealing with visual data, including the collection of videos and photos, since it is a popular choice amongst science education researchers because it reveals new nuances and insights on teaching and learning that have the potential to present better data. A quick search in Scopus (www.scopus.com) using the key words "science" AND "education" AND "video" retrieves 3920 results and the same search on ERIC (www.eric. com) returns 3770 results. For example, following the ERIC search results in one of the very first results, was an article by Siry and Martin (2014) who research preservice science teacher education. In this case, visual data has been collected to capture practices as they unfolded that were then unpacked by researchers and teachers. Other research articles take different approaches and interests in the use of video. Visual data collected by researchers and visual data produced by the participants themselves have in common the fact that they (Hulsizer 2016) represent materials that have values attached for both participants and researchers since they capture, preserve, and reveal who their participants are, and how they enact science education situations.

In the following sections I will define ethics and values with a specific reference to material value ethics by Max Scheler, to then present examples of visual data collected in science education settings that will exemplify evolving trajectories or paths that developed throughout a research encounter. The chapter will wrap up with a discussion and conclusion.

12.2   Ethics, Values and Max Scheler's Material Value Ethics

A classic definition of ethics is that ethics pertains to doing good and avoiding harm (Beauchamp and Childress 1989). When researchers prepare their research design, including their applications to ethics boards, they typically do this 'a priori', to try to foresee and address possible ethical challenges and how to deal with them. According to Max Scheler, a priori formulations are too abstract and do not address the unique responsibilities a person has for another person, since another person is not just anyone (Scheler 2014, 1954). Scheler thus objects to the Kantian position that a priori estimations made to evaluate and judge a situation are too rigid and place too much value on structure and permanence over the unique needs of a given individual or situation.

Ethical approaches are based on values that are, for Scheler, the "living moral experience in history and society" (Kelly 2011, p.4). The decisions made by science education researchers to collect certain kinds of data are thus shaped by the values of a situated research community and their practices (i.e. the living moral experience of the scientific research community specifically). Wiles et al. (2008) explained the relationship between the factors influencing ethical issues and decision-making in educational research to point out that ethical decisions are shaped by a researcher's moral framework, the research community, ethical regulations, legal regulations, professional guidelines and ethics approaches (p.8). While the authors take the moral framework of the researcher into account, what is missing from Wiles' et al. proposition are the moral frameworks of the researched and how this impacts the ethical decisions being made by the researcher and the participants. Scheler points out while all individuals have material values, they are lived, and through this process they are experienced differently by the people involved. Kelly writes about Scheler that he pointed out the need for "a sociology of knowledge to study the history and process in which values 'function' in the moral consciousness of persons and in the collective mind of a culture" (Kelly 2011, p.3–4).

Max Scheler's work is interesting because he reminds one that our ethical approaches are not so much shaped by our intellectual ability to examine things but are more often emotional acts, where we foreground some things and as things evolve, widen, or narrow our value systems change accordingly. According to Scheler, values present themselves in real material objects, actions, or people, but not as what he describes as imitations. However, through phenomenological approaches, Scheler stipulates it is possible to reflect on values also in their ideal state.

Aluwihare-Samaranayake (2012) points out that a qualitative researcher should aim to (re)present his participants and their experiences as truthfully as possible, while conducting his investigations in people's natural environments and would achieve so by directly engaging with his participants. The author writes about what it means to retell participants' experiences and the researcher's responsibilities to "represent them and their experiences in as true a form as possible" (Aluwihare-Samaranayake 2012, p.65).

This is important in order to truly understand people's social worlds and to do so would only be possible by integrating and accommodating participants' views and voices while protecting them from possible harm. This means also to critically reflect on "what constitutes socially responsible and acceptable research" (Aluwihare-Samaranayake 2012, p.75). I argue here, that Aluwihare-Samaranayake's points are echoed in Scheler's material values ethics that is about enactment of ethics.

Scheler (2014, 1954) writes that there are different levels of value modalities (in ascending order): from values of pleasure to disvalues of displeasure, from values of vitality and vital feeling, to values of the mind (truth, beauty and justice), and values of spiritual love (holy versus unholy), and finally, the values of utility versus the disvalue of uselessness. If a person would prefer a lower value it would show an incident of an emotional disorder about an earlier inclination. Scheler assumes that these kinds of value modalities are only produced if we compare different modalities and they do not represent any absolute value systems. This stance implies development, change, and transformation. When science education researchers engage with their participants in their studies, they frequently encounter changes of some sort. This could include changes in the relationship of the participants with the researchers, so adopting an ethical position that embraces the impossibility of setting everything in stone a priori, may provide a useful approach. The main focus of ethics in educational research is on how data are collected, for what purpose and if these data can be the cause of harm to individuals, or groups of people. What this means with the perspective of material value ethics in mind is discussed next.

12.3   Visual Data for Ethical Science Education Research

The performativity of ethics is intertwined with the nature of the data that is collected and the process of data collection. Early in the process, research design decisions such as how directly the researcher will interact with her participants and what kind of data is collected, shapes the nature of this partnership, including the expectations of everyone involved.

Collecting visual data to learn about the enactment of science education can be very revealing since people's practices can be captured in the context of material settings. Images or videos that show teachers and their students during science education activities reveal not only verbal but also non-verbal aspects, also shining light on how emotions play out (Kristensen and Otrel-Cass 2017; Tobin et al. 2016). However, visual data never stands alone. It usually includes a wide range of materials aside of videos and or photos. For example, transcripts from interviews or field notes from direct observations that are combined with visual data provide for rich multimodal information about people in a set context (Cowie et al. 2010). This combination of different data strengthens the validity of visual data.

Since visual data represent raw data, the process of analysis is another aspect that has to be considered in the planning of visual research since it shapes the ethics on how to handle very personal information. One of the questions to handle is over to those who will examine the visual material. Viewing videos or images with a larger research team, means that interpretations can be re-examined. Showing visual materials to the participants prompts reflexivity and helps the recall of events. Such an approach can offer new opportunities for knowledge production, create meanings and develop arguments (Pink 2013). According to Pink (2013) video ethnography, or the visual research about people's cultures, customs, habits, and differences, is about the encounter between the lived and performed life and should always aim for the production of new knowledge which is of course an aim of science education research.

After settling on a research design the process of engaging with participants typically starts with the establishment of informed consent agreement, as this allows the researcher formally to engage with them in the educational research (Heath et al. 2011). Typically, informed consent is handled in such a way that researchers start by sketching out the aim of the research to justify to their participants the need for collecting data from them. The procedures of the data collection, including dates and times are outlined before asking participants for their signature to signal their willingness to the researcher team to conduct their investigation. Informed consent should also always inform participants about their rights, in particular to be made aware that they can dissent – that is not to participate at all, or after having initially consented the right to withdraw consent again. Therefore, the researcher-participant relationship involves a form of cooperation that is based on trust (Dockett et al. 2012). To support a cooperative relationship, informed consent can be organised in such a way that participants can be given a choice a priori what data collection procedures they want to consent to and what not.

It is also important to consider who collects the data, especially when potentially vulnerable people, such as young children in schools, are involved (Heath et al. 2011), and how the data is collected e.g. is it researchers, students from the universities, or both, and whether data collection is achieved through direct observation of educational practices and through follow up analysis. To gain a deeper understanding of observations in educational settings, the data analysis could include initial reflections with experiences by and with teachers, students and the researchers after the classroom observations. Such reflective conversations in retrospect may begin with an assertion, followed by telling examples from observations or interview/discussion excerpts. These conversations seek to create knowledge production from the viewpoint of participants in an active collaboration with them (Pink 2013).

Once in the field, it is necessary that the researcher also understands the material conditions that shape particular research activities and subsequent experiences for their participants (Otrel-Cass et al. 2010) since material aspects to visual research can be significant. This means that the physical setting, including all artifacts and the tools researchers may be bringing with them, contribute to how people behave and interact. Using multiple camera angles, such as handheld and/or fixed cameras, frame the information researchers can gain about classroom interactions (Goldman et al. 2007) but may impact on the naturalness of the observed setting. In science classroom settings, it may be necessary to video record students and teachers participating in activities set in laboratories or other locations, but this may give students the impression that this is a form of performance assessment. While using more than one camera can help capture different activities taking place at the same time, it requires the willingness and the trust of the participants to provide such information. It may also require that researchers adopt a higher degree of responsiveness towards their participants' interests and an openness to incorporate what participants identify as being beneficial to them and those who they represent (Decker et al. 2011.)

If the data collection involves the collection of visual data, and since science education research involves children or teenagers, additional challenges may arise. Young people have to be able to make a judgement if they ought to be truly informed about and deeply understand the aims and purpose of a given project. Therefore, in order to gain their permission, it is typical, that up to a certain age, to obtain informed consent from disparate people such as, parents/guardians, teachers or classroom assistants (Heath et al. 2011). However, these communities hold their own situated values and may represent the children or young people's values only up to a certain degree.

Adopting a material values approach to ensure that the relationship between the researcher and participants is a trusting one, consent from participants should be about more than a 'one-off' agreement since participants live and experience the research, and their opinions and feelings about their involvement may change over time (Pink 2013; Jordan and Henderson 1995). This means that ethical considerations and negotiations should not only take place before but also *during* the process of data collection, and later on the analysis. This would be particularly important if the research aim is to gain insights from participants that requires their confidence in the research that is being conducted.

12.4   Exemplifying Enacted Material Values Ethics

The example referred to here, stems from a project called 'Beyond Technology' (www.beyondtechnology.eu), a study that was conducted in Denmark, Sweden, and Finland between 2016 and 2019. The example refers only to the Danish part of the study. The project investigated the implementation of technology at school, with a particular focus on student owned technology such as smartphones in science classrooms. The Danish case followed a class during their science lessons over the three years. With this research focus in mind it was clear from the start that the study was potentially invasive, since smartphones play a significant part in many young peoples' private lives (Devitt and Roker 2009).

12.4.1   Whose Research: Material Encounters with the Research Questions

In this first illustration I want to show what it can look like to build trust between research participants and researchers and that this means that researchers may need to adopt more flexible approaches in their research design in order to identify what can be of benefit to their participants when they agree to being involved.

In a first step, the researchers invited the teachers and their students to jointly examine the initial research proposal with the aim to explain what the ideas and concerns had been that should justify such an investigation. The activity was captured on video, photos and through observational notes for later examinations. Since the initial research proposal was written jointly by the researchers and the teachers, the team explained their ideas and hopes. Next, the students were asked to reflect on what had been presented to discuss amongst each other the research questions and then identify if other topics should be considered, topics that had perhaps more relevance to their lives. A topic that emerged during an activity where the students identified and then voted for questions relevant to them were frequently associated with their well-being, such as how smartphones disturb one's sleep, or can be used for bullying. The children suggested also ways on how to investigate those issues as part of the study.

Figure 12.1 shows some suggestions made by students who proposed to investigate what it means to be without their smartphone. We picked up on their second suggestion namely to ask some students to conduct an experiment and not to use their phones for up to a week.

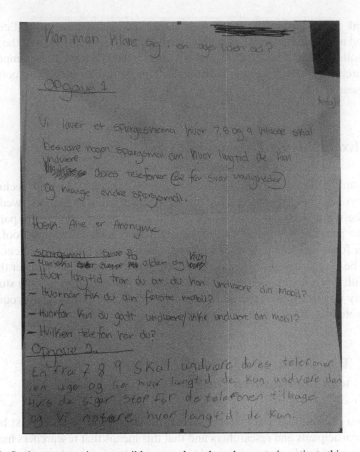

Fig. 12.1 Students suggesting a possible research angle and ways to investigate this

12.4.2 What It Means to Consent

In this next example I illustrate what it means not only to rely on consent that has been given a priori to the research that is being conducted.

Written consent was sought twice during the three years from the students and their parents. In addition, presentations and discussions about the research intentions and ambitions were conducted once a year. Parents and legal guardians were also invited to those presentations. Those meetings were video recorded (the video cameras captured only the researchers on video but recorded the voices of all who attended the meeting (with permission of all involved) in order to capture the negotiation process concerning the focus of the study between researchers and participants. The presentations included that researchers and teachers were giving verbal information and updates about the project and details about the nature of the data collection. This gave the researchers a possibility to discuss (not only in writing) what was meant by 'rights of participants including the right to withdraw from the research or parts of it' and how we could deal with this as part of the project.

Excerpts of Researcher Reflections from the Information Meetings
During the first meeting one parent asked if it was ok for their child to participate even if it did not have a smartphone. We responded that it was of course in order, specifically because we wanted to examine 'real' science classrooms.

During the second meeting one parent asked if we could tell them if we observed that their children were misusing their phones at school. We decided that it was important to protect our relationship with the students and explained to the parent that by and large we had observed that they had been always on task, which was in fact our impression. On the very few occasions when they used their phones for short non-school related activities, we asked the students about this. They would typically justified this because they had finished a task and were waiting for new instructions and needed to 'zone out' for a moment.

Conducting these meetings more than once helped to reflect on the work we were doing and built trust to collaborate as a team.

The letter of informed consent offered students a number of options to select from. We noticed that in year one a few students opted out of some of the data to be collected from them, some of which they revoked in year two. Following is a translation of the Danish original letter of informed consent (see Appendix for Danish original), showing the options students were presented with. This part was the last section of the consent form after a detailed project description at the start of the letter.

I hereby grant permission to the following (please place a cross with the items you give permission for us to collect data from you):

- Fill out a questionnaire (as described)
- Participate in interviews (as described)
- Be observed and recorded on video at school (as described)
- Be audio recorded at school (as described)
- Share videos with the researchers that I have produced (as described)

12.4.3 Becoming Materially Involved

In this example I want to highlight what it means to orchestrate data collection through the use of video cameras and what it may mean to adopt a material value ethics approach to data collection, selection and dissemination.

It was one of the intentions in this project to collaborate as closely as possible with the students. This included that we asked the students to produce videos for us. This meant that they could choreograph what they wanted to share with us and a wider community and what not. One of the students felt confident to share her video production with one of the researchers. One outcome of this development in the relationship was that this student joined the research team to present her data at an international research conference. The presentation was practiced of course but it was stressed that she should present what she felt was important to her to share. Although prominence to the students was was given the decision was made not to reveal the student's name in the published conference abstract to protect her identity (Otrel-Cass et al. 2017). This had been a difficult decision since we did not want to silence the student's contribution. We discussed this with the student and her parents. The student decided that while she felt happy to participate at the conference she preferred not to be named in the proceedings. She was orally introduced by name and affiliation during the conference presentation, so her participation was not entirely anonymous (Fig. 12.2).

Another student-led production was the preparation of a manifesto. The idea was to propose why it may be good or not to integrate mobile technology into the classroom teaching. The manifesto was prepared by the students and in iteration between teachers and researchers, as a deliberate product that resulted from the students contemplating on how they perceived the opportunities and challenges that come with an increase of technology in schools, classrooms and young people's lives. This production started with a video recorded class discussion between the researcher, the teacher and the students. In it the discussion connected back to the start of the project when the group discussed the initial research questions. The following is an excerpt from the transcript of the discussion in class:

Fig. 12.2 Researcher and student presenting at a research conference

Teacher: So, should we use smartphones more in school? Could you see any kind of use that would be good to get you to learn more or learn something different?
Student 15: Phones nowadays have a lot of sensors, like barometers, accelerometers, gyroscopes and everything... light sensors, some even have distance sensors. There a lot of apps that use those, where you can see the data that is coming of them. That could teach kids a lot about all kinds of data, acceleration, g-forces, some phones can even measure g-force.

The discussion focused on the students as expert advisers over their own lives and the technology that forms part of it. The team emphasised on several occasions that the idea was to use the discussion and the later writing of the manifesto to inform others, including teachers, parents, or school authorities. It represented an opportunity to ensure that there was a benefit resulting from the joint work for the group the students represented.

Teacher: Yes. And next year, there will be something ... about programming and coding. .....
Student: The teachers should also learn from us, yeah.
Teacher: You think the teachers must learn how to use the technology?
Student: From us.
Teacher: What about?
Student: I just mean, from us.
Teacher: From you, yes.
Student: ...because this way... it is faster.

The student here pointed out that they had expert knowledge in an area that they would be happy to share with others. The students recommended that they could act as superusers and inform younger students as well as teachers. This discussion was later transcribed and then summarised into the key points and handed back to the students for further corrections. Following this step, the students were put into groups to work on the manifesto. In a final step, the students reviewed selected video clips when they discussed these topics to reflect and revise for a final time. Interestingly, the students changed the point on students as advisers on technology for teachers since they felt that teachers ought to know about specific technological programmes that are meant for teaching and learning purposes. The manifesto and its material manifestations including the video clips represented material outputs that were developed over time and organised so that students had a chance to reflect on their own experiences during the project period and refine the recommendations they wanted to make.

12.4.4 Analysing Data – Enacting Material Value Ethics

In this last example I will present an argument how material value ethics can apply to the analysis process of research data.

The students in our project were asked to help rectify or improve the research findings. They were provided with selected video excerpts from the classroom observations. Those episodes were selected by the researchers in the first instance.

Fig. 12.3 Students reviewing video data
Student_right: (points to computer screen [00: 03: 08.27]) Ok here you can see you can use the internet to help
Student_left: It means you can combine being creative, work with your hands, and use such real things, like fossil stuff. You can then make something out of it (laughing [00: 03: 25.14]). You can use real things and combine this with the internet

When presented to the students, the researcher explained their interest in a particular episode to then continue by simply playing the short episode, a few times and then ask the students to reflect on what they thought about the selection, explain their opinion of what it showed, and if there was anything noteworthy about it (Fig. 12.3).

The key was, to hear the students' voices and adjust the interpretations stemming from the researchers. It also meant that in the conversation between the researcher and the student the researcher explained and tried to justify why they found the selected scenes intriguing which is rarely done in research. In one other instance, a student pointed out to the researcher that they had misinterpreted a scene where one student used a physical model to explain to another student a rocket rotating around the Earth. The student explained to the researcher that he used the physical model because the other student "is dyslexic and finds it hard to read the information on the screen". Through these interactions, we were able to jointly have greater insights and co-created the findings between researcher and students.

12.5 Discussion

When Max Scheler presented his ideas about material value ethics he pointed out that there was a unique obligation of the one who is doing something to another person to do this in a responsible way, but he stressed that it is about what 'I,' and not merely anyone, does (Scheler 2014, p. 94). The ethical decisions we took at different points of our investigation in the example presented here echoed our interpretations of operating with research integrity with the participants we were working with. They were not just any students to us and we tried to treat them with respect, also to produce meaningful research findings from the work in their science classroom. Scheler's discussion on ethics stands out in that it emphasises the uniqueness of people's experiences, and the different values people attribute to them. However, the value we researchers place on the information we generate from our participants is of a different nature to how they perceive their experiences.

This is particularly important since values carry positive or negative connotations and the materials that are of value may be valued positively by one and negatively by another. Thus, what Scheler explains is that values are created through intention. When we researchers pointed out to the students, their parents, and the teachers how important it is for us to make meaningful contributions about how science is being taught in modern classrooms we placed a positive value on us capturing everyday practices at school. When we explained that we wanted to collect some of our data through video collections and photos, we pointed out the risks of identifying individuals and how we could handle this risk in response to the students', parents' or teachers' concerns.

Scheler explains also that we rank the values we place from low to high, but that this ranking is an intuitive act that is anchored in the experiences we make. This was certainly the case when the parents asked us to learn more how their children managed having their smartphones in class during the science lessons. Specifically, a question emerged whether they would breach the agreement not to use social networks during school time. The material value placed on our research data that emerged here was concerning trust: the trust put in the children being allowed to use their phones during the science lesson, the trust between the researcher and the parents, and the trust between the researcher and the students. For us the relationship between researchers and students was ranked highly since we were dependent on their willingness to share their practices with us, and this trust placed in us had to be handled with care. Kelly (2011, p. 58) notes about Scheler that in his writing he "recognizes the complexity and the historical diversity of the systems of values of persons and cultures." The parents, students, teachers, and researchers in this project all represent members of nested cultures with nested value systems that have different objectives. Scheler differentiates between what he calls 'ordo amoris' or the universal order that is governed by our emotions and affections and the ethos of a community (Kelly 2011, p. 44). He writes that the ethos of a community emerges from ordo amoris, and signifies the connection between an individual and the community they belong to as well as their unique style of assigning meaning to the world. The school children represented a community with a specific ethos. This ethos is different to that of a of friends outside of school or the ethos that can be found within a family. Of significance for the science education researcher is to understand the existence of emotions and affections and the ethos of specific communities.

In the 'Beyond Technology' project the analysis process involved students and teachers' post lesson reflections to inform the analysis of sequences from the visual data produced from the classroom observations. Asking the participants to join a viewing session of video data elicited significant information about and interpretations of their perspectives on and understanding of what unfolded in science lessons that had been captured. Involving participants as co-producers of meaning and active vetting partners was an ethical decision that had methodological consequences. The idea was that this process should also strengthen the validity of what we were able to say about what we had witnessed. This approach can also be useful when certain actions are unclear to the researcher. Receiving feedback on how participants parse events can strengthen the meaning-making process and the knowledge production. It became evident in this project that when participants are involved in the social construction of knowledge they have the possibility to enact on the spot what to share and what not, thus this is how I interpret what Max Scheler describes as material values that are shaped by situated practices (Scheler 2014).

Research conducted at the micro level, that is in classrooms with students and teachers, requires constant negotiations. In this example these were taking place between the researchers, students and teachers in particular but also parents who were placing certain values in understanding science practices aided with mobile technology and other values the ways of collecting data such as through visual modes.

12.6   Conclusion

The ambition in this chapter has been to draw attention to reflective and performative ethical approaches that take material value ethics into consideration. Max Scheler's thinking has been described as a particular ethos that can be shared between the different partners within research communities. The value on collecting evidence through visual modes, such as video to research science classroom practices is thus explained as an evolving practice where values guide practice and thus become performative.

Using Scheler's notion of material value ethics can help to understand the drivers behind how research activities are conducted. The guidelines that have been produced by various research bodies (for example BERA 2018) represent a particular ethos of nested cultural groups and then transformed by researchers in their particular setting. The argument made here is that one way or another this transformation continues through the refinements made by an individual's response to research activities. In the simplest case, this could be that a participant does not consent to participate. In the example used here, steps were taken so that the individuals were able to change their ideas and also take more active roles in the science education research process.

While we as researchers aim to set a priori codes for good ethical conduct we cannot decide who a person is by giving them a universal definition. Some people do not object to being filmed while others do. Some see benefits in explaining and justifying researcher interpretations, others do not. Their interests or reluctances in collaborating are based on their histories and the values they place on different activities and why researchers need to negotiate those with care. This is what Scheler means when he talks about the element of performativity in ethics.

Taking a material value ethics approach is a reflective and responsive way to acting ethically because it identifies the individual values people hold and it "will enable us to understand the ethos of persons whose sense of good and evil is different from our own" (Kelly 2011, p.40). Utilising video data amplifies the possibilities for participants to take active roles and refine, adjust and correct what they would like researchers to communicate elsewhere.

Appendix

Danish original section from a letter of informed consent:

Samtykkeerklæring

Jeg har læst informationen i dette brev og forstår, at jeg er blevet bedt om at give tilladelse til, at der kan blive indsamlet data fra mig i dette projekt: *Beyond technology in primary schools: the role of technology ownership in different subjects and the impact on pedagogy*

Jeg giver hermed min tilladelse til at (sæt venligst **kryds** ved alle de ting **du giver tilladelse til**):

- Udfylde spørgeskemaer (som beskrevet).
- Deltage i interviews (som beskrevet).
- Blive observeret med videokameraer i skolen (som beskrevet).
- Blive lydoptaget i skolen (som beskrevet).
- Blive fotograferet i skolen (som beskrevet).
- Dele videoer, som jeg har optaget med forskerne (som beskrevet).

References

Aluwihare-Samaranayake, D. (2012). Ethics in qualitative research: A view of the participants' and researchers' world from a critical standpoint. *International Journal of Qualitative Methods, 11*(2), 64–81.

American Education Research Association (n.d.). *What is education research?* Retrieved from http://www.aera.net/About-AERA/What-is-Education-Research

British Educational Research Association [BERA]. (2018). *Ethical guidelines for educational research* (4th ed.). London: BERA. https://www.bera.ac.uk/researchers-resources/publications/ethicalguidelines-for-educational-research-2018.

Beauchamp, T. L., & Childress, J. F. (1989). *Principles of biomedical ethics* (3rd ed., p. 5). New York: Oxford University Press.

Burgess, R. G. (2005). Ethics and educational research: An introduction. In *The ethics of educational research* (pp. 10–18). London: Routledge.

Butler, J. (2011). *Bodies that matter: On the discursive limits of sex. Routledge classics*. Abingdon, Oxon ; New York, NY: Routledge.

Cowie, B., Otrel-Cass, K., & Moreland, J. (2010). Multimodal ways of eliciting students' voice. *Waikato Journal of Education, 15*(2), 81.

Decker, S. E., Naugle, A. E., Carter-Visscher, R., Bell, K., & Seifert, A. (2011). Ethical issues in research on sensitive topics: Participants' experiences of distress and benefit. *Journal of Empirical Research on Human Research Ethics, 6*(3), 55–64.

Devitt, K., & Roker, D. (2009). The role of mobile phones in family communication. *Children & Society, 23*(3), 189–202.

Dockett, S., Einarsdóttir, J., & Perry, B. (2012). Young children's decisions about research participation: Opting out. *International Journal of Early Years Education, 20*(3), 244–256.

Goldman, R., Pea, R., Barron, B., & Derry, S. J. (2007). *Video research in the learning sciences*. Mahwah: Lawrence Erlbaum Associates.

Heath, C., Hindmarsh, J., & Luff, P. (2011). *Video in qualitative research: Analysing social interaction in everyday life*. London: SAGE.

Hulsizer, H. (2016). Student-produced videos for exam review in mathematics courses. *International Journal of Research in Education and Science, 2*(2), 271–278.

Jordan, B., & Henderson, A. (1995). Interaction analysis: Foundations and practice. *Journal of the Learning Sciences, 4*(1), 39–103. https://doi.org/10.1207/s15327809jls0401_2.

Kelly, E. (2011). *Material ethics of value: Max Scheler and Nicolai Hartmann* (Vol. 203). Dordrecht: Springer Science & Business Media.

Kristensen, L. K., & Otrel-Cass, K. (2017). Emotions: Connecting with the missing body. In *Exploring emotions, aesthetics and wellbeing in science education research* (pp. 165–185). Cham: Springer.

Osborne, J., & Dillon, J. (2008). *Science education in Europe: Critical reflections* (Vol. 13). London: The Nuffield Foundation.

Otrel-Cass, K., Cowie, B., & Maguire, M. (2010). Taking video cameras into the classroom. *Waikato Journal of Education, 15*(2), 10j9.

Otrel-Cass, K., Bruun, M. H., & Gnaur, D. (2017). Primary school students as co-researchers. In The Association of Visual Pedagogy Conference (AVPC) 2017 at Aalborg University, Denmark (pp. 118–122). Dafolo Forlag A/S.

Pink, S. (2013). *Doing visual ethnography* (Vol. 3). London: SAGE.

Roth, W. M., Tobin, K., & Ritchie, S. M. (2008). Time and temporality as mediators of science learning. *Science Education, 92*(1), 115–140.

Scheler, M. (2014, 1954). *Der Formalismus in der Ethik und die materiale Wertethik*. Hamburg: Felix Meiner Verlag.

Siry, C., & Martin, S. N. (2014). Facilitating reflexivity in preservice science teacher education using video analysis and cogenerative dialogue in field-based methods courses. *Eurasia Journal of Mathematics, Science & Technology Education, 10*(5), 4.

Tobin, K., King, D., Henderson, S., Bellocchi, A., & Ritchie, S. M. (2016). Expression of emotions and physiological changes during teaching. *Cultural Studies of Science Education, 11*(3), 669–692.

Wiles, R., Prosser, J., Bagnoli, A., Clark, A., Davies, K., Holland, S., & Renold, E. (2008). *ESRC National Centre for Research Methods review paper: Visual ethics: Ethical issues in visual research*. National Centre for Research Methods.

Kathrin Otrel-Cass, is a Professor of education and digital transformations at the University of Graz, Austria. Her research interests are often of interdisciplinary in nature with focus on digital visual anthropology and variety of qualitative, ethnographic methodologies. She works with various practitioners and experts in environments where people are working with science/technology/engineering practices or their knowledge products. Her research is often set in schools but is not exclusive to those environments. Her research interest in visual ethnography has led to the establishment of a video research laboratories at Aalborg University and the University of Graz with a focus on the organized analysis of video recorded data and ethical research practices involving visual data. Kathrin is also a member of the Human Factor in Digital Transformation research network at the University of Graz.

Chapter 13
Methodological Ethics Considerations in Science Education Research: Symmetric, Authentic, Material, Adaptive and Multidisciplinary

Martin Riopel

13.1 Introduction

Following the previous chapters focusing on methodological considerations related the ethics with research in and about science education, this afterword aims to synthetize and reflect on some common or related issues raised in these chapters. It will first propose two dimensions for which the focus of the filed is shifting: the sharing of responsibility and the invasiveness of data. It will then conclude that these shifts are signs of maturity in the field and show some alignment, or parallel developments, with developments in science education research and current society.

13.1.1 A Shift of Responsibility Toward More Complex and more Equitable Relationships

The first important dimension discussed by the authors is the relation between researchers, participants and stakeholders for which it is proposed that the balance of responsibility and benefits could be shifting away from the researcher to create more complex and more equitable relationships.

In their chapter, Andrée et al. (2020) insist on the importance of symmetry in participatory science education research. This symmetry principle applies primarily to teacher-researcher collaboration for the teaching interventions but can also be expanded to ontological, epistemological and methodological values commitments at play. The proposed shift toward symmetry between teacher and researcher could produce more engagement of teachers in research activities as well as more

M. Riopel (✉)
Université du Québec à Montréal, Montréal, Canada
e-mail: riopel.martin@uqam.ca

involvement of researchers in teaching practices. This greater collaboration should eventually lead to more appropriate responses to problems related to teaching practice both in theoretical and concrete ways.

It is interesting to note here that this symmetry principle applied to teacher-researcher interactions can also be considered for the teacher-student interactions as proposed by Tabak and Baumgartner (2004) for inquiry-based science learning and that research suggests that this leads to more pedagogical efficacy. This parallelism between the recent evolutions of ethics related to science education and the evolution of science education itself is not really surprising and may be interpreted as an indication that something more general is going on.

In another chapter, Adams and Siry (2020) focus on the importance of living authenticity in science education research and propose how to enact an authenticity criteria that extends the usual ethics considerations to encompass all stakeholders and recognize subjectivity and context-dependent structures that mediate research outcomes. This aims to increase the potential benefits from their experiences, from a subjective and contextual point of view, while they are participating in a research project. This proposed shift toward authenticity could produce research projects with more understandable outcomes but also make individuals more informed, more understanding of others, more stimulated by research and more engaged toward change.

These outcomes related to applying authenticity to research in science education are not unlike those related to applying authenticity to science education (Braund and Reiss 2006; Roth 2012) and even to other disciplines (Rule 2006). Once again, some parallelism can be observed between recent evolutions of ethics related to science education and the evolution of science education itself.

Pushing to encompass even more, Scantlebury and Milne (2020) propose a post humanistic approach to ethics that includes all non-human material entities when questioning education research practices, methods, data analysis and interpretations. Science education research with this approach is intrinsically contextual, dynamic, relational and should take into account generally how humans and matter are entangled in knowledge production but also more specifically how teachers, students, researchers, material settings and instruments are entangled in producing learning science activities. This proposed shift toward materiality could produce more complete descriptions and understandings of what matters in science education research processes. In this context, as human and non-human entities become more and more entangled, equilibrium between benefits and risks in ethical decisions has to change by considering, for example, that researchers are always part of the experiment and cannot make completely external decisions or that responsibility could be shared in some cases with machines with the corresponding risks.

Once again, this evolution of ethics about research in science education can also be considered for science education itself. For example, Snaza et al. (2014) propose to apply post humanism to education by considering the entanglement of humans and technologies that is becoming more and more important. This could eventually lead to a fundamental learning displacement from human to human-machine, human-animal or even human-machine-animal collaborations. This is a major shift that somehow encompasses the all the more equitable propositions of precedent chapters in the sense that researcher, participant and stakeholder are entangled in their relation to all physical realities and beings and that all relative responsibilities should be acknowledged accordingly.

13.1.2 A Shift of Focus Toward Continuous Micro-Level Data

Another important dimension discussed by the authors is the extent and the potential invasiveness of micro-level data to be continuously collected, mostly visual, that makes the researcher a privileged witness that should also develop a greater ethical respect of participants' privacy. In some cases, new types of data such as those related to neuroscience raise issues that even require a development of ethics.

In their chapter, Hilppö and Stevens (2020) insist of the students' ethical agency in video research. They propose that the rapid development of various recording technologies has created new opportunities that redefine the conventional boundary marking what people can know about each other in a way that accentuates the researchers' obligations of respectful and diligent treatment of this knowledge, especially in research with children. To address this concern they show how children indicate their awareness of the audience and create privates spaces for interactions not to be recorded. Ethical symmetry in this context commands to conduct research the same way with adults or with children and consequently to explicitly and respectfully renegotiate boundaries of the research. This shift toward invasiveness of technologies balanced by awareness of children's implicit choices could lead to a continuous renegotiation of consent which is a major methodological challenge related to ethics but also a major challenge to society where technologies are also becoming more and more invasive.

Afterwards, Otrel-Cass (2020) uses a material ethics approach in visual science education research. She insists on the fact that, while researchers aim to set a priori codes for good ethical conduct, they cannot decide who a person is by giving them a universal definition. Some people do not object to being filmed while others do. Some see benefits in explaining and justifying researcher interpretations, others do not. Their interests or reluctances in collaborating are based on their histories and the values they place on different activities and researchers need to negotiate those with care. Ethical considerations in this context cannot be fixed a priori and should adapt continuously to individuals and to situations. As pointed out by the author, this shift toward adaptability and responsiveness related to ethics with potentially invasive visual data could produce globally more trust: trust between researcher and students, trust between researcher and parents, and even trust between children and parents.

Although the shift towards micro-level data from video-ethnograhies has important consequences in terms of ethics, there are also other developments of data production in science education research challenging practices of ethical reflection, in particular that of neuroscience. These psychophysiological and neuroscience data are more private and more sensible because they have unique and profound ways for peering into the body and into the brain (Shamoo 2010) and they raise some ethical issues.

First, even if it has been usually recognized that neuroscience research can contribute to the science education field (Ansari et al. 2012; Masson and Brault Foisy 2014; Masson et al. 2014; Blanchette Sarrasin et al. 2018; Smyrnaiou et al. 2016; Riopel and Smyrnaiou 2016), this is still a subject of passionate debate (Bruer 2006; Horvath and Donoghue 2016). This fundamental contribution of neuroeducation is of course necessary for the corresponding data to be useful at all. These possible benefits have to be ethically balanced with the possible risks. For example, there has been concerns related to dangers of misinterpreting or misusing the corresponding findings because of their highly technical and confusing nature (Alferink and Farmer-Dougan 2010). Another ethical issue is the invasiveness of such technologies that also needs to be considered very seriously. As pointed out by Ansari et al. (2012), recent availability of non-invasive methods to image the brain reduces these risks and makes it possible to measure school-taught skills in authentic contexts.

Considering more generally ethical approaches to neuroeducational research, Howard-Jones and Fenton (2012) propose an interdisciplinary stance that focuses on three main areas. The first area concerns conducting research at the interface of cognitive neuroscience and education. The comparison between the two fields leads to propose that physical risks are mostly alike but that psychological risks differ slightly in the area of incidental findings where educational researchers don't always have the same expertise. More differences are observed when comparing the social and educational risks because the participants are usually more engaged and their voice is usually more heard in educational research. Consequently, because of its high complexity, neuroeducation could lead to less implication of participants than other educational research and this social risk has to be ethically balanced with the possible benefits. The second area of ethical issues to neuroeducational research is interpretation and communication of findings. The entanglement of many disciplines in the neuroeducation field makes it difficult not to propagate neuromyths and other misuses. To avoid them, it is proposed that higher quality standards and interdisciplinary expert collaboration could be ethically required for research communication in neuroeducation. The third area is policy making for which many emerging and difficult ethical issues have been identified such as cognitive enhancers, neural infant screening and genetic profiling. These cannot be resolved with existing set of ethical principles from contributing disciplines and will require more interdisciplinary expert discussion and more public consultation and debate.

13.2 Alignment with some Challenges of Current Society

All the precedent propositions focusing on methodological considerations related ethical issues with research in and about science education can be interpreted primarily as a sign of maturity in the field: they all extend research ethics from a more classical and general stance (macro-level, researcher-centered, prior consent) and simple unilateral relationship to a more actual and specific stance (micro-level, participant-centered, continuous renegotiation, multidisciplinary) and equitable relationship. This can be viewed as reassuring from a non-specialist point of view as it implies an evolution toward more awareness and sensitivity from the researchers' community and leads to decentralization and sharing of power and responsibility. This evolution of methodology ethics in science education research is in alignment with the evolution of science education itself. These propositions are also in alignment with challenges of the current society as they try to address, in the context of respectful research ethics, the important issues related to the omnipresence and invasiveness of technologies in individuals' life. This invasiveness can be observed with data from computers and phones, social media and videos, but also from psychophysiological and neuroscience data that are linked to the emerging field of neuroeducation. One can hope that these new ethical issues will lead to fruitful interdisciplinary discussions and collaborations and that these could in turn serve as inspiration to public consultation and fruitful debate.

References

Alferink, L. A., & Farmer-Dougan, V. (2010). Brain-(not) based education: Dangers of misunderstanding and misapplication of neuroscience research. *Exceptionality, 18*(1), 42–52.

Andrée, M., Danckwardt-Lillieström, K., & Wiblom, J. (2020). Ethical challenges of symmetry in participatory science education research. In K. Otrel-Cass, M. Andrée, & M. Ryu (Eds.), *Examining research ethics in contemporary science education research*. New York: Springer.

Ansari, D., De Smedt, B., & Grabner, R. H. (2012). Neuroeducation–a critical overview of an emerging field. *Neuroethics, 5*(2), 105–117.

Blanchette Sarrasin, J., Nenciovic, L., Brault Foisy, L.-M., Allaire-Duquette, G., Riopel, M., & Masson, S. (2018). Effects of teaching the concept of neuroplasticity to induce a growth mindset on motivation, achievement, and brain activity: A meta-analysis. *Trends in Neuroscience and Education, 12*, 22–31.

Braund, M., & Reiss, M. (2006). Towards a more authentic science curriculum: The contribution of out-of-school learning. *International Journal of Science Education, 28*(12), 1373–1388.

Bruer, J. T. (2006). On the implications of neuroscience research for science teaching and learning: Are there any? A skeptical theme and variations: The primacy of psychology in science learning. *CBE-Life Science Education, 5*, 104–110.

Hilppö, J., & Stevens, R. (2020). Northwestern University, United States Students' ethical agency in video research. In K. Otrel-Cass, M. Andrée, & M. Ryu (Eds.), *Examining research ethics in contemporary science education research*. New York: Springer.

Horvath, J. C., & Donoghue, G. M. (2016). A bridge too far–revisited: Reframing Bruer's neuroeducation argument for modern science of learning practitioners. *Frontiers in Psychology, 7*, 377.

Howard-Jones, P. A., & Fenton, K. D. (2012). The need for interdisciplinary dialogue in developing ethical approaches to neuroeducational research. *Neuroethics, 5*(2), 119–134.

Masson, S., & Brault Foisy, L. M. (2014). Fundamental concepts bridging education and the brain. *McGill Journal of Education, 49*(2), 501–512.

Masson, S., Potvin, P., Riopel, M., & Brault-Foisy, L.-M. (2014). Differences in brain activation between novices and experts in science during a task involving a common misconception in electricity. *Mind, Brain and Education, 8*(1), 44–55.

Otrel-Cass, K. (2020). The performativity of ethics in visual science education research: Using a material ethics approach. In K. Otrel-Cass, M. Andrée, & M. Ryu (Eds.), *Examining research ethics in contemporary science education research*. New York: Springer.

Riopel, M., & Smyrnaiou, Z. (2016). *New developments in science and technology education*. New York: Springer. 203 p.

Roth, W. M. (2012). *Authentic school science: Knowing and learning in open-inquiry science laboratories*. Springer Science & Business Media.

Rule, A. C. (2006). The components of authentic learning. *Journal of Authentic Learning, 3*(1), 1–10.

Scantlebury, K., & Milne, C. (2020). Beyond dichotomies/binaries: 21st century post humanities ethics for science education using a Baradian perspective. In K. Otrel-Cass, M. Andrée, & M. Ryu (Eds.), *Examining research ethics in contemporary science education research*. New York: Springer.

Shamoo, A. E. (2010). Ethical and regulatory challenges in psychophysiology and neuroscience-based technology for determining behavior. *Accountability in Research, 17*(1), 8–29.

Siry, C., & Adams, J. D. (2020). Living authenticity in science education research. In K. Otrel-Cass, M. Andrée, & M. Ryu (Eds.), *Examining research ethics in contemporary science education research*. New York: Springer.

Smyrnaiou, Z., Riopel, M., & Sotiriou, M. (2016). *Recent advances in science and technology education, ranging from modern pedagogies to Neuroeducation and assessment* (p. 390). Newcastle upon Tyne: Cambridge Scholars Publishing.

Snaza, N., Appelbaum, P., Bayne, S., Carlson, D., Morris, M., Rotas, N., Sandlin, J., Wallin, J., & Weaver, J. A. (2014). Toward a posthuman education. *Journal of Curriculum Theorizing, 30*(2), 39–55.

Tabak, I., & Baumgartner, E. (2004). The teacher as partner: Exploring participant structures, symmetry, and identity work in scaffolding. *Cognition and Instruction, 22*(4), 393–429.

Martin Riopel (Ph. D) is a professor of science and technology education and vice-dean of research at the Université du Québec à Montréal (UQAM) of Canada. He was also a member of the institutional research ethics commitee for many years. His research interests focus on computer-assisted learning, serious games, learning models and neuroeducation. He also holds the research chair on educational innovation at the Paris-Saclay University of France.

Index

© Springer Nature Switzerland AG 2020
K. Otrel-Cass et al. (eds.), *Examining Ethics in Contemporary Science
Education Research*, Cultural Studies of Science Education 20,
https://doi.org/10.1007/978-3-030-50921-7

Printed in the United States
by Baker & Taylor Publisher Services